Deep Space Warfare

Deep Space Warfare

Military Strategy Beyond Orbit

JOHN C. WRIGHT

McFarland & Company, Inc., Publishers
Jefferson, North Carolina

Library of Congress Cataloguing-in-Publication Data

Names: Wright, John C., 1983– author.
Title: Deep Space Warfare : Military Strategy Beyond Orbit / John C. Wright.
Description: Jefferson, North Carolina : McFarland & Company, Inc.,
Publishers, 2020 | Includes bibliographical references and index.
Identifiers: LCCN 2019043201 | ISBN 9781476679266 (paperback :
acid free paper) ♾ | ISBN 9781476637846 (ebook)
Subjects: LCSH: Space warfare. | Outer space—Strategic aspects.
Classification: LCC UG1530 .W75 2020 | DDC 358/.84—dc23
LC record available at https://lccn.loc.gov/2019043201

British Library cataloguing data are available

ISBN (print) 978-1-4766-7926-6
ISBN (ebook) 978-1-4766-3784-6

Front cover image by Anton Chernigovskii (Shutterstock)

Printed in the United States of America

McFarland & Company, Inc., Publishers
Box 611, Jefferson, North Carolina 28640
www.mcfarlandpub.com

To the giants on whose shoulders we all stand

Acknowledgments

This book would not have been possible without a host of individuals and groups who contributed immeasurably to my education. First, I am immensely grateful to the faculty and library staff at Air University, Air Command and Staff College for their tremendous advice, interpretations, and sound wisdom concerning this work. Major Brent Ziarnick, USAFR, was essential to this work's completion by providing solid feedback and lending me his extensive space knowledge, especially regarding space's influence on economics. Colonel M.V. "Coyote" Smith, USAF (ret.), was always willing to listen and to help refine my broader thoughts on space and its applicability to international relations. USAF Lieutenant Colonel Pete Garrison's healthy obsession with space attack provided excellent inspiration for this work's writing on space dominance and planetary invasion. Major Sam Kriegler, U.S. Army, offered me his superb insight and useful sanity checks for the manuscript, not to mention a critical eye from a sister service with comparatively little space responsibilities. Finally, the United States Air Force must be thanked, not only for giving me the opportunity to meet these tremendous scholars and warriors, but for the various opportunities I have been given which have made me a better officer and thinker, and ultimately led to this work.

I must also thank the legions of friends and colleagues who encouraged me along the way to continue writing, despite the awkwardness which always comes with writing about a subject with which one is not at first completely familiar. As a combat pilot, writing about space was intimidating both because there is so much to learn about the subject, and because I know there are multitudes of space professionals ready and willing to pounce on my conclusions and make short work of them. Nevertheless, I owe them my gratitude as well, not only for their critical eye but also for their willingness to stand up and fight for the right mix of military strategy, policy, and technology to meet the nation's imminent space challenges.

Lastly, I have profound gratitude for my family. They can never be adequately repaid for the time and inspiration they have given me. I was able to complete this work only through their encouragement and support.

I have done my best to represent space, military strategy, and all that comes with it as accurately and honestly as I could, but no doubt there will be misrepresentations and mistakes found in this work with such a contentious topic. All errors or misinterpretations in this volume are mine.

Table of Contents

Preface

In 1911, the airplane was in its infancy. Before the Great War broke out, a few military airmen were able to explore this wondrous new technology, but most ordinary people were acquainted with the air via the tame and well-established hot air balloon. For those who stumbled upon a hot air balloon ride, many became entranced by the calm smoothness of a basket-encapsulated ascent, the sight of their home from the air, and were no doubt surprised at the unexpectedly gentle travel afforded by light breezes, further placated by the stillness that comes without engine noise. Flight seemed more like a holiday than a weapon. As the airplane became militarized, as nearly all technology inevitably is, this harmonious environment gave way to a third dimension for the modern battlefield, replete with fervent competition for the high ground via aircraft that flew higher, faster, and with as much weaponry as could be carried. A final plea for the air's innocence was given on the eve of World War I by the English novelist John Galsworthy, who implored the military via the *London Times* in 1911:

> If ever men presented a spectacle of sheer inanity it is now—when, having at long last triumphed in their struggle to subordinate to their welfare the unconquered element, they have straightway commenced to defile that element, so heroically mastered, by filling it with engines of destruction.... Is there any thinker alive watching this utterly preventable calamity without horror and despair? Horror at what must come of it if not promptly stopped; despair that men can be so blind, so hopelessly and childishly the slaves of their own marvellous inventive powers.... For the love of the sun, and the stars, and the blue sky, that have given us all our aspirations since the beginning of time, let us leave the air to innocence![1]

His pleas, while appealing to humanity's reason, fell short of humanity's ambition.

There is an analogue between the airplane and the spacecraft. As of this writing, the spacecraft is the primary dominion of the government. It exists only to prove to our species that we can get to space and use it; nothing more. Everything else that space technology has accomplished thus far, including the so-far brief and rudimentary exploration of our solar system, the best telescopes our species has produced, and even our nascent satellite network around this planet, all exist to better our lives on the surface; to watch, intimidate, or destroy our enemies; or, in the very near future, to carry exceptionally rich tourists to lower earth orbit. In the end, however, space will be weaponized. At that time there may be an outspoken advocate like Mr. Galsworthy, but the seduction of the high ground, regardless of form, will always attract military strategists—and their innovations. While the time for planetary combat is not yet upon us, interplanetary and interstellar warfare's future form is worth examining.

This book was written to seriously examine the military consequences of humanity's eventual weaponization and warfare activities in deep space. It was written for audiences in military, academic, and policy spheres, as well as citizens interested in realistically examining space warfare away from the diversionary pleasures of science fiction. It is a product of the author's imagination, military experience and love of science and space, and borrows heavily from his education as an Air Force officer. It is the author's hope that this book, and others which will surely follow, will encourage national policymakers, legislators, and military professionals to view space as a warfighting domain which will shortly be fully weaponized, and which is deserving of special attention before formulating appropriate space policy and strategy. Above all, it aims to convince this audience space warfare is closer than we think, and consider the implications of being unready to tackle its challenges in a sober manner divorced of the traditional derision that comes with thinking about space warfare. While it attempts to address seriously the future military problems sure to be found during our interstellar adventures, it was also enjoyable and fun to write.

Our species may not agree on the form space warfare beyond our own planet will take, but we can all agree humanity will eventually, unstoppably, enter space and do so aggressively, safely, and repeatedly, just as we did with the air.

Introduction: War Plan Orange

> A good Navy is not a provocation to war. It is the surest guaranty of peace.
>
> —President Theodore Roosevelt,
> Address to the Congress, 2 December 1902

Neither Japan nor the United States wished to go to war in the 20th century. In 1906, each state respected the other's accomplishments in their particular sphere of the world. President Theodore Roosevelt, who had just negotiated the Treaty of Portsmouth in 1905 between Japan and Russia, was "thoroughly well pleased" by Japanese victory over their mutual rival.[1] Japan, for its part, was busy building its administration over its newly acquired territories, including holdings in Manchuria and a major presence in Korea which would lead to annexation in 1910. Relations with the United States were amicable, and both states shared interests in the overall peace and stability of East Asia. Suddenly in April of 1906, Californians looking for an easy scapegoat following a major earthquake in San Francisco stirred up trouble by blaming and attacking "Orientals" in the state, including Japanese immigrants.[2] Some Californian politicians went so far as to restrict property rights of Japanese immigrants and segregated Japanese schoolchildren from whites, which enraged rising Japan.[3] Feeling pressure from the Japanese government, the Roosevelt administration was eager to defuse the temporary "war scare," and knew American jingoism when they saw it. In all, both sides thought little of the temporary outburst and order was soon restored with no major damage done to U.S.-Japan relations.

While cooler heads prevailed this time, the brief incident forced both Japan and the United States to seriously think about the possibility of war. Given the technology and geopolitics of the day, neither nation seriously expected nor believed war was feasible. Nevertheless, a group of U.S. naval officers at the Naval War College began cobbling together a series of ideas which would eventually become War Plan Orange, a strategy designed to conduct a maritime campaign against Imperial Japan.[4] The plan designated the Japan side as Orange, and the U.S. side as Blue as part of a series of international war plans in which potential combatant states were assigned colors for ease of planning.

The plan was eerily predictive of what would eventually happen in 1941. In Plan Orange, U.S. planners predicted Japan would strike the first blow as swiftly and as decisively as they could, targeting U.S. naval assets and military posts throughout Asia, without going through the trouble of first declaring war.[5] Legendary naval strategist and U.S. Navy Captain Alfred Mahan predicted Japan would mimic its initial assault on Russia in the Russo-Japanese War of 1901–1902 as its main blueprint for attacking the United States. Planners were in general agreement that Japan's objective would be to push the United States out of the Western Pacific by depriving it of its territories and allies, then holding out as long as possible during an extend war of attrition until the United States was forced to sue for peace due to domestic political exhaustion.[6]

Given the lack of urgency surrounding Plan Orange, U.S. planners enjoyed a fairly free hand in their musings about what Orange forces could or would do, and what Blue forces could do about it. As both Japan and the United States would again determine in the 1940s, U.S. planners in the 1910s already understood U.S. national wealth could support an indefinite naval campaign.[7] Similarly, both understood Japan, if unleashed, would fight until the last man or until ordered to stop. The only remaining unknown was how long U.S. public support would last, which was impossible to predict. These problems remained on December 7, 1941, after Japan had made its decision to seize U.S. and European holdings in the Pacific and to enlist time to whittle down U.S. forces and U.S. public support for the war now thrust upon them.[8] Japan underestimated U.S. popular resolve; the United States underestimated just how much destruction Japan would endure until surrender.

The plan was one of the most prophetic in U.S. military history. This is more remarkable in that it was planned off-and-on over the course of 35 years from 1906 to 1941, and did not include the actual campaigns of World War II. In its first of three phases, Plan Orange assumed Japan would conquer U.S. assets in the western Pacific and then go on to claim the vital raw materials and oil further south.[9] The plan called for a sacrificial holding action in these U.S.-held territories, which included Guam, the Philippines, Wake Island, and potentially even Midway, in order to allow the U.S. fleet as much time as possible to marshal its fleet from Pearl Harbor and begin steaming west.[10] The plan assumed the U.S. Navy had free action at its facilities in the eastern Pacific, especially Hawaii and San Diego, from which the second phase would launch.

In the second phase, scheduled to occur about six months after Phase I, Blue naval forces would sally in force to Japanese waters in the western Pacific, fighting Orange vessels the entire way, with the aim to reclaim its lost naval bases in the Philippines. Surprisingly, this phase of Orange was planned to last two or three years, and was understood by U.S. planners to primarily involve naval attrition as Japanese forces were expected to attempt to trade small holding naval actions for time. The plan also clearly identified the possibility for greater attrition in waters closer to Orange supply lines with their greater Orange fleet concentration.[11]

In the third and final phase of Plan Orange, Blue forces would advance from their newly re-established bases in the Philippines and strike at Orange vessels and Japanese home territory with both naval and airborne assets (air power was added in later iterations of the plan). This offensive action was to continue until Japan sued for peace, thoroughly convinced of U.S. determination to end the war and persuaded U.S. public support would not buckle.[12]

To be sure, the plan was not perfect. While great detail was given to cruise distances for maritime movements, as the years went by less attention was paid to available aircraft and facilities, much less to airborne distances to the Japanese home islands. Further, a "glaring deficiency" of the plan was a U.S. inability to deploy large aircraft to the Pacific within striking distance of Japanese holdings in the plan's target areas, especially the Philippine islands.[13] In the 1930s and 1940s, aircraft and aircraft carrier technology clearly obliged militaries across the world to secure land-based runways and facilities as fast as possible to support larger bombers and strike aircraft. Truly destructive aircraft required runways much longer than carriers could provide to conduct operations. Where Orange aircraft would be a short hop from their forward-deployed and well-established air bases, Blue aircraft would somehow have to cross the Pacific safely first before they could drop a single bomb.[14]

Nevertheless, this more-than-one-hundred-year-old plan has profound implications for interplanetary and interstellar warfare. Plan Orange was conceived at first as a "just in case" war exercise addressing a potential foe the United States never thought it would have to fight. It addressed an enemy separated from U.S. territory by thousands of miles over tyrannically empty Pacific waters, and demanded an unknown but certainly high number of vessels, equipment, and sailors to traverse this waste, contested the whole way by both enemy action and the environment. Blue forces were unsure of what they would find when they arrived at the Philippines, but knew only their relative technological and materiel superiority would eventually provide the weight to crack the shell of any fortification or defense effort Japan could offer. Despite the clear knowledge of this risk, naval planners were cognizant the plan would take several years to successfully execute, and acknowledged the plan's relatively high attrition rates. In short, it was a dangerous campaign planned with little knowledge which was to be conducted over a great distance with a largely unknown enemy at a higher than expected cost.

Space will offer us much in the same way as Plan Orange. Regardless of the foe, regardless of the destination, war plans for space operations will resemble Plan Orange in the titanic distances and logistical planning required to successfully execute them. As explained in this book, space combat by its nature demands more logistical support than terrestrial operations, more planning, and is less forgiving than terrestrial war. The environment itself opposes any force sent into the great void between planets or star systems; distances between objects *within* a single system are significant enough to present an extensive menu of risk and death. Like Plan Orange, space operations war plans will need to be aggressively revised as technology continues to advance. Like Plan Orange, political requirements will shape space war plans and drive their imperatives.

In one regard, through science fiction humanity has already offered itself a glimpse into what such plans could look like, and what warfare in space could one day become. While primarily created for entertainment, the host of creative artists and writers which have produced science fiction have led us down several fantastical roads into the inky blackness of space. While certainly many science fiction works and films are completely forgettable, and some so ridiculous they are better left forgotten, a notable cadre of visionaries have made profound observations about human existence, advanced technology, and issues which humanity will surely face someday in space.

While it is impossible to list all of the most influential science fiction creators here, some of the most prominent and believable ideas add value to the discussions in this book. Robert Heinlein tackled the issues of high technology, planetary settlements, military service and government, factional warfare, and a host of other issues in his landmark work *Starship Troopers*. Isaac Asimov brought robotics and ethics into the casual conversations of people across the world with his groundbreaking *I, Robot*, along with the disquieting consequences which come with crafting sentient beings to serve other sentient beings. Jules Verne, the father of modern science fiction writing, took readers on fantastic journeys in works such as *20,000 Leagues Under the Sea* and *From the Earth to the Moon* with unthinkable technology at the time of writing; his contraptions are now either being actively used by humanity or are becoming more and more believable. H.G.

Wells in *The War of the Worlds* brought us the terrifying images of an alien invasion into our homes, first broaching the topic of a cosmic attack—and one in which humanity loses—to the casual radio listener. William Gibson's *Neuromancer* series grappled with future societal breakdowns due to technological proliferation and the darker side of computers, networks, and global information.

Film and television have contributed as well. George Lucas' *Star Wars* needs no introduction, and its classic space adventure is many a youth's first exposure to science fiction. Gene Roddenberry's *Star Trek* addressed terrestrial problems like international politics, equality, culture, diplomacy, and war in a space setting, as well as presenting believable technology and real scientific principles beneath a thin science fiction veneer. A series of fascinating and well-produced films have allowed humanity to explore the more imaginary side of space, including *2001: A Space Odyssey*, *Alien*, *Event Horizon*, *Edge of Tomorrow*, *Planet of the Apes*, and more recently *The Martian*. In particular, this last film's emphasis on real scientific problems, including what it would actually entail to survive on Mars and how help is truly far away given the distances found in space, captivated thousands and was one of the most popular films in the world the year it was released. It would seem a healthy dose of realism in a science fiction work is a powerfully imaginative catalyst for a public which grows steadily more scientifically literate and readier to accept space operations as part of modern existence.

While most people are tangentially familiar with space warfare due to these works, this book does not endorse them. This book seeks to address humanity's eventual decision to pursue war and the use of force in a deep space setting, and the military strategic consequences of doing so. Since humanity has made war in every other domain—land, air, sea, and now cyberspace—it is safe to assume deep space is not far behind. Like these other domains, humanity will eventually find a way to regularly traverse space and conduct safe, repeatable, and dependable military operations within its starry blackness. Unlike earlier eras, talking about combat in space is no longer seen as absurd and time-wasting sojourns into fiction, akin to how a child pretends they are traveling to the moon in a cardboard box. Rather, when space conversations come up today in military conversations, the topic is discussed with seriousness and attention. Military commanders listen when their space officers—whose very existence speaks volumes—present their concerns. People no longer find spacegoing battle fleets and armed astronauts far-fetched. Science fiction is by far the most "acceptably nerdy" interest to possess amongst a general slice of the population, not only because it represents windows into the future, but because it is now a socially acceptable and genuinely interesting topic which many people can apply to their daily lives and the general technological progress around them. While we cannot build a future on science fiction, we can thank it for its contribution to conditioning the masses to space and its possibilities.

When humanity is ready, its militaries will need to craft reliable, flexible, and dependable operations plans, like Plan Orange, to achieve security objectives sure to be threatened by those wishing to seize the commanding view provided by space. These plans must soberly examine space's vast distances, logistical troubles, and must seriously consider the form combat in space will take. Before discussing the proper strategic approach to interplanetary and interstellar warfare, one must first understand the basics behind war in the stars, and what lessons our wars here on Earth can teach us about how humanity will approach space warfare.

1

Interstellar Basics

> Two possibilities exist: either we are alone in the universe, or we're not. Both are equally terrifying.
>
> —Arthur C. Clarke

When most people think of space warfare, they do so from a whimsical perspective of "alien versus human." In reality, space warfare is already here, albeit currently restricted to orbital reaches near and around Earth. In fact, the most difficult thing for an average observer to imagine is a *weaponized* space domain, where actual state-run vessels and machines fight force-on-force battles several hundred kilometers above the surface of the Earth. Extrapolating from this reality, interplanetary and interstellar space warfare will come much sooner than humanity will encounter other sentient life capable of competing with us on an interplanetary scale.

To be sure, the first enemy humanity will encounter in space, and the first it already has, are other humans. As of this writing, the state is the premier user of power in space, and it is therefore reasonable to assume state-on-state conflict will be the first kind of open space warfare we as a species will experience. Aside from science fiction film and literature excursions, most people do not think about interstellar or interplanetary conflict on a daily basis. For starters, there's no reason to do so until a threat emerges. For another, it's in many ways *unnatural* to spend our precious time and energy on low-probability conflicts, which is better spent on physical and social survival. For the same reasons the United States doesn't worry about being invaded by The Maldives, regular people do not fear rival states' spacefaring war vessels, nor sinister extraterrestrial influences.

Nevertheless, it is foolhardy, unscientific, and unsound strategically to assume we are the only sentient species in the universe, and therefore, it is unwise to plan as though we are. Thus, it is instructive to consider the very real biological possibilities of encountering another sentient spacefaring competitor. Many discussions of spacefaring extraterrestrial life begin by discussing the Drake Equation, a calculation proffered by Cornell astronomer Frank Drake in 1961. The equation itself is subject to numerous criticisms and is usually calculated simplistically in order to foster discussion or due to a lack of actual data.[1] The equation is

$$N = R^{*} f_p n_e f_l f_i f_c L$$

where N is the number of civilizations in the Milky Way capable of emitting detectable electromagnetic emissions, R is an average of the number of stars born each year (accepted as 1.5), f_p is the percentage of star systems around which planets form (about 90 percent according to observations, or 0.9), n_e is the number of Earthlike planets on average per solar system (estimated to be 1.5), f_l as the fraction of these planets which actually develop life (displayed as 1 for simplicity's sake, meaning all Earthlike planets develop life), f_i are the ratio of planets which birth life and then develop intelligent life (usually represented as two-thirds, or 0.67), f_c is the percentage possibility of planets with intelligent life developing civilizations, and finally L as the likely length of time civilizations will remain detectable before they die out, advanced so far technologically as to no longer be detectable by Earth technology, or some other calamity which caused their electromagnetic emission capability to disappear.[2] For those interested, a casual calculation yields approximately 340 spacefaring civilizations remain to be detected during the lifetime of our civilization. Only a slight change in one or two variables results in drastically different results, making the Drake Equation conjecture at best.

Thus, the equation is designed to shave down successively the number of planets where a civilization could reach sufficient development to communicate with Earth. While the equation itself is worthy of its own discussion, it is immaterial to this text.

Conflict

Humanity has evolved, grown, and developed around conflict. This conflict, more often than not, has meant violence and the use of force. Our capacity to wage war and to easily access the darker aspects of our nature is a subject which has been studied extensively, and is fundamental to understanding the human condition and our eventual conduct during interplanetary and interstellar conflicts.

Historian John Keegan once put it clearly: "warfare is almost as old as man himself, and reaches into the most secret places of the human heart, places where self dissolves rational purpose, where pride reigns, where emotion is paramount, where instinct is king."[3] War as we know it has been around at least 12,000 years. In its wake, no culture, no nation, no period of history has been untouched by its ghastly hand.[4] It is accepted as part of human nature, and as a necessary growth factor in civilization; after all, without war and conflict, humanity never would have felt the competitive forces which pushed technology further, made walls higher, made weapons sharper, and forced people to think harder in order to guarantee their survival.[5]

Space Conflict and International Relations

The basis of all conflict between political organizations—which is any group larger than an individual—lie in interactions between what international relations scholars Joseph Nye and David Welch call the primary political units.[6] Conflict occurs because these primary political units representing different groups use power to better ensure their survival, and this use of power affects different political units unevenly. There are many ways to use power, but this book is concerned with military force, a key instrument of national power which attempts to coerce other political units to conform to a preferable behavior which enhances another group's ability to survive. Humanity has always reserved the right for itself to use violence; the only difference when comparing the use of violence throughout history is often the size or type of political unit which claims this right. Before organized governments and the modern state, the primary political unit for humans was often the family or tribe. As human collectives grew larger and more sophisticated, so did the need to more strictly control the use of force and harness its power for more deliberate and focused political use. Since we are talking about the use of force as power, "political" is the correct descriptor for this activity.

The right of political units to exercise violence on behalf of their political objectives is known in international relations as the use of force. For most on the planet today the right to exercise violence in pursuit of political goals is claimed by the state.[7] This right is sourced directly from primate behavior and humanity's nature as a species; in this regard, the use of force carries with it the designations "lawful" or "unlawful," depending on whether or not legitimate political units use this right. This legality is established by common understanding and culture rather than a rule or written law that somehow applies to the entire species. The use of force has seldom been successfully challenged philosophically, and its use to defend organizations is taken for granted as one of—if not the principal—obligations of the primary political unit. The tribal chief orders a raid on a rival, for example; in another, the central government's military protects us from attacks. The best arguments against a "lawful" use of force humanity has developed have yet to convince anyone this right is somehow obsolete, unnecessary, or truly illegal. The Kellogg-Briand Pact of 1928, for instance, was signed by fourteen countries largely hailing from the industrialized world.[8] In it, the signatories formally renounced the use of force as a method to resolve international disagreements. While a noble effort, the signatories could not be convinced to give up their rights to self-defense, and therefore to the maintenance of armed forces. The U.S. Senate, before ratifying the treaty by a wide margin, did so with the caveats that the United States reserved the right to protect itself militarily and to "act against signatories breaking the agreement"—which of course meant going to war to punish treaty offenders.[9] The Kellogg-Briand signatories would go on to butcher each other on battlefields across the globe less than twenty years later.

Today's dominant primary political unit is the state. This term "state" can be confusing, depending on the context in which it is used. It can mean a country, or a government, or even a sub-sovereign level of political authority (like in the United States and Mexico, where the immediate political level below the federal level is a "state," as in the State of Missouri, and so on).[10] Despite this limit of language, the meaning of state is clear in one regard: a state retains absolute control over a fixed territory on Earth. This control is called sovereignty. In space, we can expect territoriality and sovereignty to someday apply to areas, planetoids, stellar bodies, or any other items of interest claimed by states, and these claims will cause conflict. While the Outer Space Treaty of 1967 currently prohibits states from claiming territory in space, in reality only political power can enforce or reject that claim; only conflict can make the final decision.[11] As we shall see, the distances between stellar objects and the logistical difficulties in supplying space forces means a preponderance of military power in one area or near a particular object of interest will make that force the de facto arbiter of that object's security and access.

States today function in an anarchical environment. This does not mean states operate in absolute chaos; rather, this means there is no overarching political entity *above* the state to impose its will upon sovereign actors.[12] In other words, there is no overarching authority states can appeal to for help against other states. While the United Nations (UN) can enact resolutions and demand states take action, it cannot reliably (as yet) enforce these prescriptions through military power or other forms of coercion. While the consequences for ignoring UN decisions are notable, only other states can truly impose UN rulings upon other states. For example, the United States–led coalition which liberated Kuwait from an Iraqi invasion in 1991 was conducted under the auspices of a United Nations resolution. Since the United Nations lacked a world-class military to enforce its decision, the United States and its allies were obliged to do so. As such, the international "system," referring to the anarchical environment, is often termed a self-help system.[13] This simply means when the primary political units want something, they must get it for themselves; they cannot rely on a higher authority to appeal to with their request, nor on other states to give up something of theirs. Trust is a serious issue in such a system, where no state has a monopoly on force, and a state pursuing what it needs could eventually (and often does) lead to war.[14] Suspicion and wariness are very normal conditions in the international system.

In the event all states existing on Earth one day politically unify—an extremely unlikely prospect and discussed further in Chapter 3—these facts will not disappear. Rather, sovereignty would merely expand to encompass the entire planet and human holdings in space. Someone must make the political decisions, no matter the size of the primary political unit, and disagreements about how we should distribute the wealth and power of human political organizations will always challenge the notions of wide-scale political unity.

Why Space Will Be an Anarchical Environment

Space's unique characteristics suggest our current anarchical international system will continue as human political units spread throughout the stars. Space is comprised of extreme distances in between relatively unfamiliar stellar bodies, and those bodies are by and large hostile to human life. Any territorial claims made in space will therefore likely occur with an implicit promise that a primary political unit who holds the interest in them must maintain, supply, and manage these claims. Because the international system is a self-help system, claimant states cannot rely on other political units to safeguard these claims for them. Moreover, because of space's unique environment, territorial claims in space will share three major characteristics: they will require technological sophistication, they will be complicated, and they will be vulnerable. All three have major implications as to why space will continue to function as an anarchical environment for rival political units.

First, territorial claims made in space will require technological sophistication. This is because the environment, deep space, is hostile to habitation and permanence, and because logistical support is difficult to provide and often distant. Consider, for example, a scientific research station orbiting a stellar phenomenon. Not only does this station require sophisticated and sensitive scientific equipment to accomplish its mission, but it also requires everything the onboard personnel (presuming there are any) need for their daily existence in between resupply efforts. A political entity is presumably profiting from the discoveries made at this station; this entity, therefore, has the obligation to provide the station's logistical support, as well as its security. This in turn requires a sophisticated resupply and/or maintenance system, which may require distant travel through the hostile blackness of space. How many resupply or maintenance vessels would be required to service such a station? How many should be sent in case one does not arrive safely, is late, or is lost enroute? Do these vessels require military escort? What kind of supplies are necessary? Has adequate communication between the station and its home government provided the station the chance to inform their superiors what exactly it is they need? Is there a communication system in place which satisfactorily proves to all parties that adequate and clear communication has even been accomplished? Each of these efforts by itself is a hefty lift for political organizations which have never done them before, and space's unique environment means each and every territorial claim, be they planetwide settlements or tiny mining stations, will also have unique needs depending on their local environments. Safeguarding, maintaining, and protecting these claims requires thought and foresight. It requires sophistication.

Territorial claims will also be complicated. While technological claims like the above research station example are surely complicated from a technical standpoint, physical territorial claims in space are another matter entirely. Unlike land, stellar phenomena and area in space are three dimensional. This means primary political units which claim large swaths of space will most likely tend to bite off much more than they can chew—or protect. On paper, claiming a particular stellar body, planet, moon, or nebula looks deceptively simple. But in reality, square kilometers suddenly become *cubed* kilometers, literally exponentially increasing the amount of territory the unit claims—and must subsequently defend. Disagreements about how stellar bodies' boundaries are determined, such as how far off a planet's surface a territory extends, are the clear political minefields of our spacefaring future. This problem of territoriality is tailor-made to produce political rancor. Imagine, for example, that a particular state claims an entire nebula for itself. A rival state, who also has interests in the nebula, must therefore either ask for access or contest the first state's claim. Does the claim encompass internationally agreed-upon standards? How does the claimant measure the nebula's boundaries? Can the claim be proven? If the claim is ignored, does the claimant state make a credible threat of force to defend their claim?

To make matters worse, stellar phenomena, unlike Earth territory, are usually in motion in the cosmos. This presents a great deal of complications to how humanity thinks about territory. Paris, for example, is always there; its latitude and longitude are set, and one who travels there today will find it in the same spot several years from now. One's favorite café will probably be on the same corner tomorrow as it was today. In space, though, planets, asteroids, comets, and all manner of stellar phenomena move according to their type of body and gravity's influence on them. Planets circle their sun or suns with fairly predictable patterns; but comets swirl around stars in a more carefree manner, often with huge periods between when they come near a particular reference point. Haley's Comet, for example, is only visible by those on Earth once about every 75 years. Is it claimable? If so, would a state commit resources to find it and chase it down while on its tremendous journey? Nebulae from our previous example are simply collections of gas; this gas, by its nature, is not beholden to any boundary. How could this be claimed? Even the galaxies themselves are moving and spreading out towards each other. Space territorial claims are complex and challenge human though about what territory actually is.

Third, claims in space are vulnerable. This is because of two main characteristics which space shares with the maritime domain: space as a medium is difficult to traverse repeatedly with pinpoint precision, and the relative distance between territorial claims combined with relatively slow speeds of spacegoing vessels means territorial claims will by their nature be left unprotected more often than not. Political units will no doubt place defenses and garrisons at or near territorial claims which are truly important; planetary settlements, strategic resource mines, trade routes, or others. Nevertheless, no political unit will ever have enough military or security forces to truly police the myriad possible claims in and around the entirety of a political unit's claimed territory. Imagine, in our previous example when a state claimed that nebula for itself, that a rival state has been discovered in a distant part of the nebula's confines. Its presence violates the first state's claims. Presuming this interloper is not interested in ceasing its violation on its own volition, the claimant state has two choices: do nothing, or coerce the rival state from using the nebula. If the decision to coerce the state is made, this will probably involve policing the nebula with military force. Given the nebula could encompass several million *cubed* kilometers of space to search and protect, any effort to enforce a political unit's claim this way could instantly sap the entire resources of any spacefaring security force. This means difficult choices will have to be made by future space forces, similar to choices made by today's maritime forces here on Earth. Unlike on Earth, though, distant territorial claims in space could take months or years to traverse. This fact further complicates security decisions a political unit must make in defending its vulnerable territorial claims.

Some scholars point out anarchy is not the only form of international system which can exist. One form, a hegemonic empire, is another possibility. This requires complete and total sovereignty over the totality of the international system, whatever that happens to be.[15] On Earth, this would mean complete dominion over the entire surface of the planet and its political goings-on, which has never been achieved in human history and is likely never going to be. A second possible form for the international system, a feudal system, is simply a hierarchical system of political obligations which are not limited by territory.[16] While this is certainly possible, it is highly unlikely since humanity has already agreed upon territoriality as the main principle underwriting sovereignty. Some can make an argument that future planetary settlements, mining establishments, or other inhabited human interests in space could benefit from some form of quasi-independence because a great distance will separate them from the political reach of their parent governments. However, it is unlikely a spacefaring government would expend vast resources and bear the tremendous logistical and economic risks required to establish these places, and then subsequently allow them an unhelpful level of political autonomy. This phenomenon has already been learned during the West's colonial period from the 17th to the 20th centuries. For most of these colonial powers, politically neglecting their colonies taught them some hard lessons about regional political autonomy. Given the above, an anarchical system is the most likely system to prevail in space, and as such military entities can begin planning for this reality now.

Determining Military Objectives in Deep Space Warfare

Going to war in space, as with any terrestrial conflict, must be clear in its objectives and aggressive in its pursuit. Deep space's harsh environment immediately pits any military operation, of any size, against the clock. Objectives will be discussed more in-depth in Chapter 5.

Before discussing objectives, a strategist must consider the risks of placing forces in space to begin with. Given the time, distance, and treasure invested into any spaceborne force, interstellar warfare carries special penalties for the unprepared, the hothead, and the poor strategist. Due to space's nature and the astronomical distances at work in such a domain, unlike terrestrial forces, when spaceborne forces are launched they are not easily recalled or returned, nor are they easily reinforced. This limits margins for error and adds extra pain to military planners and leaders who have not done their homework.

Though it may seem distant, wondrous, and unattainable, space is fundamentally no different than any other warfighting domain. It is simply a place with unique physical characteristics, a medium like water or air, which presents specific challenges to military operations. These challenges can be overcome with technology and determination, and can be done so repeatedly and safely.

Any political entity, state or non-state, will go to war in space first and foremost to achieve a political objective. When a political entity is a state, these objectives fall under the collective term "national interest."[17] Since national interests differ between states, conflict is the inevitable result of states' interests clashing with each other. This should not be a surprise in an anarchical international environment, and we should expect to see this result as our political objectives expand to encompass interests in space.

When states decide to use military power to secure an interest, the most important thing they can do is clearly define the national objectives which the military can achieve. U.S. Air Force Colonel Edward Mann explains states must calculate a few things first before unleashing the dogs of war. First, once a state has chosen an objective, military power must be capable of achieving the goal or producing the desired result, often called an "end state."[18] The term "end state" is a misnomer; war is simply an action along a continuum of international relations interactions, and in truth international relations have no discernable "end." Next, once objectives are assigned to the military, they must be achievable with the amount and type of forces available to the state.[19] Finally, military objectives must possess the means by which they can measure their own success. This requires honest and effective standards of measurement, beyond body counts and bombs dropped.[20]

Once political objectives have been clearly delineated and a state has committed the use of force to achieve it, the military can begin planning how it will achieve the state's desired "end state." (See Figure 1.)

This planning is only a smaller subset of the overall political environment. Once a plan is developed, it must be executed swiftly. Objectives must be meticulously followed by military authorities, or else risk the possibility that objectives might create themselves later on within the fog, friction, and

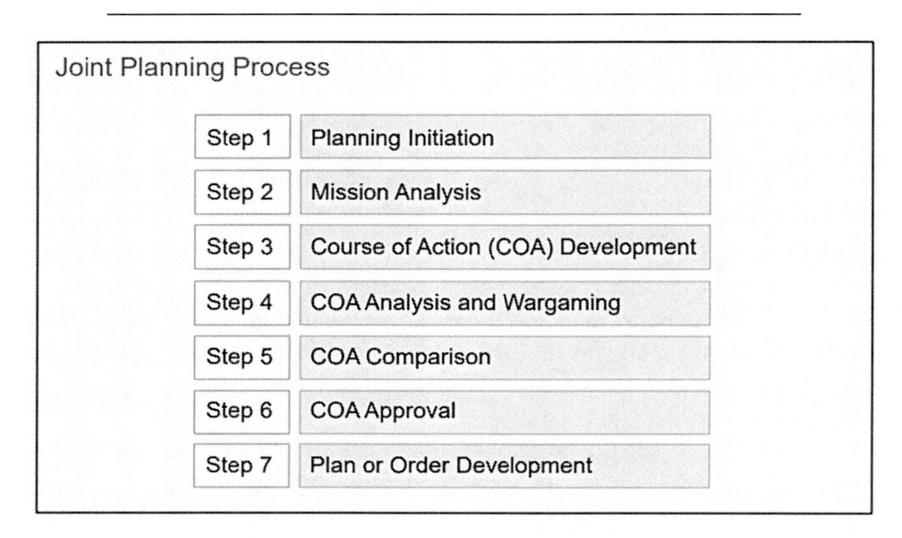

Figure 1: Joint operation planning process steps (U.S. Military Joint Doctrine, Joint Publication 5–0, "Joint Planning," p.V-2 [2017]).

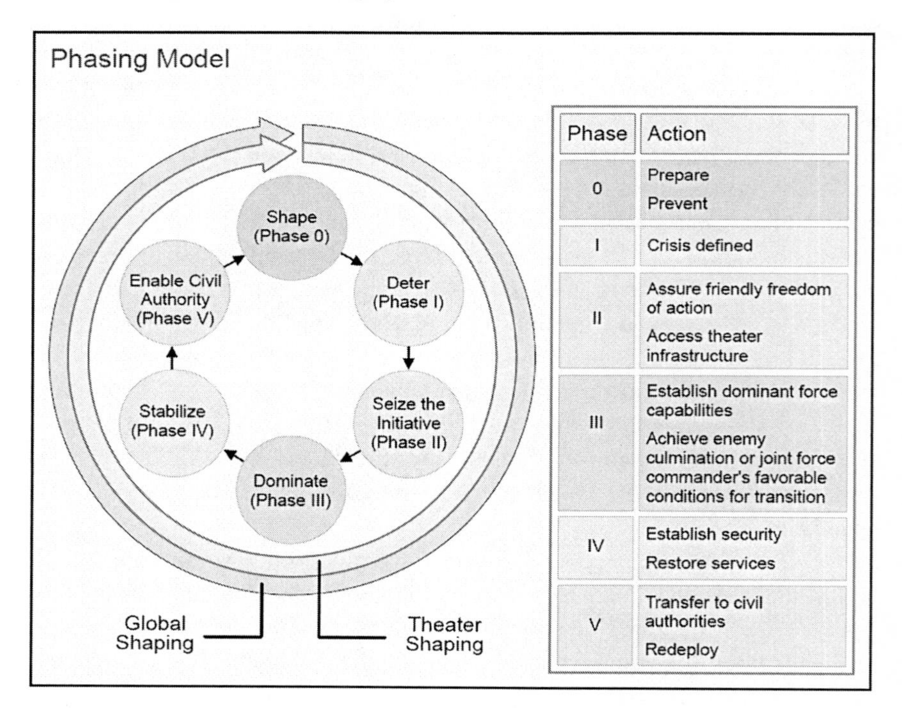

Figure 2: U.S. joint warfighting phase model of warfare (U.S. DOD Joint Publication 5–0, p. III-41 [2011]).

frustration of battle.[21] The U.S. military views its role as temporary in any conflict, and intentionally plans to carry on a conflict only long enough to relinquish control and rebuilding efforts to civilian authorities. The role the U.S. military played in this transition to civilian authority until recently is shown in Figure 2.

The cycle depicted in this chart does not restrict military planners to this structure; rather, the six-phase model merely provides a general roadmap describing how a state pursues national interests by using military force, as viewed from a military perspective. As noted above, this political objective-end state cycle is not truly endless; particularly in phases 0 and V of the chart, military activity is at a minimum and the state pursues its national interests with relatively little use of military force. Notably, this way of thinking about war is unique to U.S. military doctrine and not necessarily shared by every state, which can cause confusion and miscommunication between rivals who see things differently during conflict. Indeed, this thinking is always in flux; in U.S. military joint planning documents published after 2017, this six-phase model can no longer be found in print, though the phases themselves are retained as useful planning tools. This reflects the reality of changing military planning processes through our experiences in war; and our experiences planning military objectives here on Earth will be very applicable to identifying and pursuing military objectives in space.

The Tyranny of Distance

Space is big. Very big. Current estimates as of this writing vary, and defy our puny measurement capabilities which are much more calibrated for everyday measurements than for universal scales. Even describing a galaxy's basic dimensions defy measurement; the best way we can describe our local galactic neighborhood is only via distances between stars, and by listing the quantity of the stars themselves. That is the best our brains can do; the sheer numbers challenge human calculations, and are by no means intuitive in any way. One estimate states the Milky Way galaxy, our home galaxy, contains about 200 billion stars—a safe enough number for one's sanity—but quickly becomes incomprehensible when the estimate swells to 300 sextillion for the entire universe.[22]

To further confound our computations, the National Air and Space Administration (NASA) led a team analyzing data from a spacecraft launched in 2001 called the Wilkinson Microwave Anisotropy Probe (WMAP).[23] This probe had one mission: isolate and study the oldest light in the universe, still detectable in background radiation left over from the Big Bang. The WMAP team announced in a landmark 2013 report that the mission was a success, and revealed some startling revelations. First, by all measurable data, the universe is flat.[24] While stars certainly cluster in three dimensions, they appear to cluster on a more-or-less straight plane. Second, the study appears to have confirmed most of space—95 percent by the report's estimate—is what we would consider to be empty space, containing mostly "dark matter" and "dark energy." The nature of dark matter and energy are vociferously and enthusiastically studied and debated, and the discovery of what they entail will no doubt have great impacts on our future activities in space. This 95 percent, of course, is not truly empty; as deGrasse Tyson notes, there are a multitude of fascinating planetoids, comets, dust, gas, and matter of all kinds to be found in low concentrations throughout this 95 percent.[25] This makes sense when one considers how matter was distributed in the universe after the Big Bang. But from a military perspective, this "empty space" for now means darkness, coordination problems, silence, communication difficulties, supply concerns, and a great deal of operational risk. For the purposes of this book, the empty 95 percent has no military value at all.

One of the first scientists to seriously attempt to educate the public at large about space was the legendary American science educator Carl Sagan. On his landmark television series *Cosmos*, Sagan said it best: "The size and age of the Cosmos are beyond ordinary human understanding, lost somewhere between immensity and eternity is our tiny planetary home. In a cos-

Figure 3: The strategic map[26] (European Space Agency website: http://sci.esa.int/gaia/60192-gaia-creates-richest-star-map-of-our-galaxy-and-beyond/).

mic perspective, most human concerns seem insignificant, even petty. And yet our species is young and curious and brave and shows much promise."[27]

The distances associated with space flight are mind-numbingly challenging; traveling at light speed (which we cannot currently do), it would take four years to reach our closest neighboring star, Alpha Centauri. Military forces conducting campaigns in space will require complete re-tooling, or exceptionally fast space flight, to prepare for movements comprising years in duration.

To make matters worse, there are precious few things which can mitigate the effects of distance on a force in motion, and most of them are expensive. These include carrying additional supplies and replacement parts, which slow travel; journey slower, which could expend more fuel and expose a force to a hazardous environment longer; establish multiple logistical hubs along the travel route; and find new navigational routes to the destination. Aside from these methods, chance plays a pivotal role in seeing a military force safely to a destination intact. Therefore, a safe method for large-scale (e.g., fleet-sized) deep space travel, along with developing repeatable, reliable procedures, will be critical to any force aiming to traverse the galaxy.

Given the distances between stars in deep space, and the universe's reported 84-billion light-year diameter, it is no wonder there has never been any reliable evidence indicating any contact with any sentient extraterrestrial entity of any kind, anywhere on Earth. This fact includes radio waves or signals, travelling at light speed, which could indicate intelligence. The distances are simply too formidable; equipping and deploying a force of any kind is even more so.

The Stellar System: The Territorial Unit of Interstellar Warfare

In approaching deep space from a military perspective, because space is 95 percent (militarily) empty, one must ask what a military objective would look like. Would it be a floating military space station, constructed to remain in a stationary location, vigilantly on guard? What about a travel corridor of some kind; a natural phenomenon which strategically strangles movement between stars or black holes? Perhaps the "interplanetary superhighways" that represent the shortest distance between two stars?

A look at terrestrial naval theory provides some insight. As terrestrial navies switched from sail to steam in the 19th century, several prominent naval theorists attempted to advocate for how navies should be used. The two main schools of thought which appeared were strategies favored by Captain Alfred F. Mahan, U.S. Navy, and Sir Julian Corbett, a British attorney and maritime enthusiast. Mahan, whose influence and ideas by far reached the most ears during his lifetime and whose influence continues today, advocated an offensive use of naval forces for the purposes of destroying an enemy's commerce and seeking pitched battle upon the high seas.[28] As he was heavily influenced by legendary British Admiral Horatio Nelson, whose own personal policy was to seek and destroy the enemy wherever they were to be found, this is no surprise. Corbett supported a concept which created a "fleet in being"; that a naval force by its very nature and existence was best used as a deterrent. Corbett concluded that naval forces cannot be everywhere at once due to the vast nature of the ocean. If a navy sortied to seek pitched battle, if it cannot find the enemy fleet all it has accomplished is to waste fuel and failed to aid the war effort of its parent nation. Corbett thought a navy should either conduct operations which aided ground forces and affect the enemy's landlocked operations—such as through blockade—or otherwise disperse itself until it became possible to muster in one spot to give battle.[29]

We already know that space is 95 percent empty. From a military and materialistic perspective, this means deep space is 95 percent useless. Presuming military objectives would be to control or safeguard things which matter to our species, this means a portion of the remaining 5 percent which we deem critical to our interests automatically becomes the military objective. Of this remaining 5 percent, the stellar system or star system is clearly the most critical stellar phenomenon to our future interests in space and should represent the basic territorial unit in space campaigns. A stellar system is defined as a star or stars (optionally referred to as a system's "sun" or "suns") and the stellar bodies, planets, and planetoids which orbit that star. It is important to note that there are many more stellar systems which consist of binary star systems—possessing two suns—than there are stellar systems with one sun. In this regard, our solar system appears to be rare.

From a military perspective, star systems contain nearly everything we as a species (and presumably other similar sentient species) care about and need to continue activities in space. This includes planets, energy and mineral resources hidden inside asteroids or other planetary bodies orbiting stars, water (locked in ice), and fuel resources—some of which we have certainly not yet discovered. A military presence in a stellar system enables forward supply stations, which can lead to scientific exploration, resource exploitation, and a more secure military presence. Gravity in star systems is measurable, stable, and predictable. Furthermore, since most star systems contain at least two stars, the interplay between these suns and their orbiting planets creates the potential for electromagnetic clandestine activity. Concealment strategies in space will be discussed in a later chapter.

The All-or-Nothing Character of Space Warfare

When in trouble or overcome in battle, space forces, much like naval and airborne ones, will tend to be "all or nothing"; they will win and live, or will lose and die. The reasons for this are many, and begin with the nature of operating in a hostile environment not suited to humanity. Clearly the slightest mechanical or life support problem could lead to an entire vessel succumbing to the cold vacuum of space. In addition, there are two aspects of modern warfare which were discovered during modern terrestrial naval combat which applies to spaceborne forces.

The first is that naval technological advantages tend to be decisive and unstoppable against obsolete naval technology. First made abundantly clear by the onset of "ships of the line" and maritime vessel "classes" in 17th- and 18th-century European navies, vessel characteristics like size, number of guns, hull strength, and speed gave more advanced vessels near invincibility over smaller and less equipped opponents. Old guns fail to penetrate more advanced hulls; smaller vessels, while nimble, may not be able to maintain the same speed as larger ones due to smaller sails, weaker engines, or less efficient power plants. Smaller and cheaper vessels possess less gun mounts and likely host older technology like navigation and targeting systems. This means the older and more obsolete a vessel is compared to another combatant, the worse the potential combat consequences are for the older ship.

The second aspect is that once a naval force begins to press its weight upon its opposition and it begins to gain the upper hand, the destruction of the weaker force begins to cascade at an exponential rate. The issue of cascading destruction is, like naval and aerial combat, especially pertinent to space combat. Rescue and personnel recovery in space is unlikely since the very environment is hostile to all life, and because the enemy remains in a battle area in a relatively larger concentration as the loser begins to lose even more vessels. This means as battles in space progress, it is important to understand that the longer a forlorn fight goes on, the more the loser stands to lose.

Cascading Destruction: A Lesson for Space Combat

Both these factors will be present in any space fleet because deep space so closely resembles the maritime environment. Interestingly, these characteristics also apply to air-to-air combat today in many ways, albeit air combat is faster and often more prone to chance and less predictable than combat on the high seas.

This is true even for a force which is matched evenly, and is best understood by considering a modern aerial example. In our example, a group of twenty friendly fighter aircraft and a group of twenty enemy fighter aircraft encounter each other in the skies. For the purpose of this example, each group is armed with modern weaponry and fourth-generation fighter technology of comparable sophistication. Military pilots often train to this scenario, and each discussion before training starts the same way: what will make the difference between them and us? If we're the same technologically, will chance alone help one side prevail? As it turns out, this question has been answered many times throughout history in actual combat, and the result is always the same: unless they are at a significant technological disadvantage, the pilots who possess the best combination of training, experience, and teamwork win every single time.[30]

In the above air-to-air engagement example, there are three distinct phases which are applicable to future space warfare. The first is contact; in this phase, two groups of opposing forces discover each other, or one discovers the other first, and position themselves to fire standoff weapons (usually in the form of radar guided and heat seeking air-to-air missiles). Survivors proceed to the next phase: classic fighter maneuvering, also known as dogfighting. In this phase, fighters fly past each other physically, called "the merge" amongst military pilots, and begin to single out target aircraft. After this, they maneuver themselves into positions of advantage and kill the enemy. In any phase, the less advantaged force begins to crumble as soon as they begin taking casualties. As the better pilots or those armed with the best technology begin to kill or disable enemy aircraft, a victorious pilot is free to focus his efforts on another enemy (or double-team one enemy) as the opposing force continues to lose aircraft. One can see the carnage that quickly develops for the defeated side of an aerial engagement; in fact, the main factor which saves a defeated force from being wiped out completely during an aerial defeat is the lack of remaining weaponry available (remaining missiles and so on) which invariably occurs as a fight drags on and weapons are expended during hits and misses.

Classic naval surface engagements resembled this model of cascading destruction as well, and came long before fighter engagements. Once a naval force begins to prevail over the other, the gunners always find new targets to turn their cannons upon; torpedoes are seldom wasted on sinking or disabled vessels. Spaceborne forces will suddenly find themselves forced back into a previous era where capital ships blast each other without mercy. Indeed, the losers of future ship-on-ship space engagements could conceivably be obliterated to a ship, with the loss of an entire fleet in a single engagement. This makes space combat riskier, and will forces leaders to commit forces less often, especially without overwhelming superiority or in desperate need.

A naval ship, once sunk on Earth, at least has a chance of saving some of its complement as they bob in the waves. Rescue in space, though, is leaps and bounds more difficult than rescue under a blue and welcoming oxygen-filled sky. Recovering the remnants of a space fleet after a defeat may not be an option given a preponderance of an enemy force remaining in a battle area. This naturally shackles space forces to the same problems which naval forces have wrestled with since their inception: provisioning, lengths of voyages, reliable crews, strength and loyalty of the officer corps and the chain of command, quality of training, and the most stubborn worry of them all, the eternal specter of vessels potentially being outclassed by an enemy's unknown capabilities.

The Perilous Nature of First Contact

When one considers the scale at which the universe exists, compounded with the unstoppable onslaught of scientific progress and our species' slow reach for the stars, it becomes patently obvious that if an extraterrestrial sentient competitor exists somewhere, it is only a matter of time before we make contact. This claim is contentious, but from a military strategy perspective it is reasonable enough to contemplate potential military confrontations and associated countermeasures. The first contact with a non-human sentient being is termed, appropriately enough, a "first contact scenario," and carries with it special unpredictability and danger. While we do not yet know what a first contact scenario would look like, there are a number of military realities we can extrapolate from what we already know to be true concerning competition between species and groups here on Earth.

Deep in the South Pacific on the island of New Guinea, Jared Diamond relates a curious cultural dilemma that was once very common amongst our species. According to Diamond, who has conducted extensive research on the island of New Guinea, tribesmen out traveling alone through the jungle occasionally bump into each other on the trail.[31] Knowing full well they are from rival tribes, and always traveling armed, the two men are immediately faced with a choice: do they fight, or not? Both are completely alone amidst the creeping vines and dense atmosphere of a thick, encroaching jungle; both are cognizant of the fact that there is no friendly face to run to for help should they encounter trouble, and both know there is no one to witness and tell any tales should one end up murdering the other. In short, it is a perfect security dilemma.

Both men immediately make attempts to size up the other: they begin to judge their potential opponent's relative size to their own, their equipment, armament, demeanor, physical condition, and mental condition; both look for any weakness to exploit if necessary. After this admittedly icy beginning, an unusual event begins to take place: the men begin speaking to one other, attempting to determine—or manufacture—a reason they should *not* kill the other. After exchanging greetings, both men begin to share with the other their long list of relatives common to both their tribes, hoping to find a commonly known individual they both know via a mutual personal relationship, which they can then use as an excuse to refrain from fighting.[32] In a fascinating cultural development, New Guineans look for ways their tribes have comingled in the past to prevent conflict, murder, and mayhem now.[33]

While we can learn many lessons from this once-common exchange, the seemingly pleasant diplomacy between two men in the jungle searching for commonalities to prevent them from murdering each other hides a darker fact: they both know if they can find no common ground, one man will likely perish, and the fight itself will eventually be uncovered and war between the two tribes will engulf them and their families. In other words, the diplomacy doesn't occur because of some noble aspect of New Guinean culture. Rather, it happens to prevent a more hideous possibility—an uncontrollable wave of violence.

Does the lesson of the New Guineans doom us to similar conflicts in space, especially between humanity and a non-human species, should we encounter it one day? No; but it does reveal how humans will behave as we always have. Knowing our own reaction to this type of situation could mean the difference between peaceful interstellar coexistence, or runaway interplanetary warfare.

First contact scenarios, situations in which two or more groups with no formal contact encounter each other for the first time, deserve special mention. Of all the events which could shape the relationship between two groups or two sentient species, first contact is undoubtedly in the top three. There is perhaps no riskier venture for either party than to reveal the nature, presence, or interests of one civilization to the other. The very presence of an alien intelligence, especially if it is the first one ever encountered by one of the civilizations, has serious ramifications on the progress and character of a civilization. The primary reason for this is the first universal need of all civilizations: security and safety amidst a universe riddled with danger, many of which are of our own making, most of which are out of our control. Any species considering to intentionally enter an overt (as opposed to covert) first contact scenario should carefully consider its goals and intentions, which are not limited to but include the following.

- What military intelligence is available on the projected target species? What is known from observation, and is this observation already known to the target?
- What are the interests of both parties? If planned beforehand, to what end should first contact be initiated?
- If intentional, what is the reason for initiating first contact? Depending on the reason, who should initiate the contact? For instance, conquest (via military preemptive attack), discovery and curiosity (via scientific envoys), or diplomatic interchange (via government representatives)?
- How should first contact be initiated? What would present the least obvious threat, if we are intending to avoid behaving threateningly?
- What if the other species contacts us first? What precautions should we take? Who should meet the envoys, and how much security is actually required?
- Is communication possible? How does each species communicate, if known?
- What are the probable reactions of each species' population?
- How much study has been done, if any, regarding the counterpart species?

This is just the tip of the iceberg. Every single one of these considerations, and the many not listed here, carry a risk of failure and misunderstanding, which usually leads to miscommunication, confusion, panic, and potentially open conflict or war. Due to what could go wrong, as opposed to what little could go smoothly, the military professional must maintain a pessimistic view of first contact situations and assume the worst. Without a common background and culture, and without distant relatives to have in common, panic and misunderstanding which could lead to war is a very prominent possibility during encounters in the cosmic jungle.

After these risks, there is one final fact which inclines initial contacts to conflict. There is no guarantee, nor reason, that any other human group, or any other sentient species, encountered elsewhere in the cosmos can or should value, recognize, understand, or think about any concepts which another group understands in a similar way. War, however, is a universal: it is as clear as the light from a red dwarf when someone else is trying to kill you. When it comes to conflict, the subject of this book, the only inter-species consideration is scale. A single force-on-force skirmish may not degenerate into open war depending on the perspective of each species; the more willing each is to avoid conflict, the better the chances of buying time to develop better communication, a closer working relationship, or even ways to mitigate the political impact of accidental firefights. Indeed, fellow sentient species are also aware of and can commit overreactions, accidental weapon discharges, and panicky military moves just as well as we can.

We should also remember that until two sentient species have established a clear, correct, and dependable method of communication, both sides can essentially "get away" with fighting the other, with no recourse available in the form of diplomatic parlay. A lack of communication between two sentient species can only mean a perpetual Cold War, which always brings with it the potential for heat. This concept will be discussed in a later chapter.

Cautiousness begets military preparedness. Any party to a first contact situation should, above all, be circumspect and armed as a precautionary measure. For the human species, whether or not an individual or a group is considered a threat is usually determined based on a track record of positive, peaceful relations. From our perspective, we must understand we will naturally view any non-human civilization as an untrustworthy unknown quantity. Knowing our reaction in this regard will help plan our initial diplomatic overtones to the unknown species.

First contact is an obviously volatile situation. When it comes to force, though, the chance for violence is multiplied by the clear odds that any first contact situation will be one-sided in terms of military power. This point is emphasized by the fact that there are only two types of first contact situations: planned and unplanned. If the first contact is planned, one side will likely sortie with a force greater than their counterpart to ensure their survival and to intimidate a potential rival in order to deter future aggression. This policy, while it can be perceived as aggressive and "rude," is in fact the safest course of action a power initiating contact can take. In this case, the force initiating first contact—and it will most certainly be a government-led and funded military expedition to ensure maximum safety and success—has by definition surveilled the extraterrestrial culture and prepared a proper introduction. In this case, sortieing without a preponderance of force would be foolhardy, even at risk of appearing to an extraterrestrial civilization first and foremost as a military bully. If it were humanity conducting the first contact, there will be plenty of time to communicate the benefits and accomplishments of human culture *after* first contact and initial diplomatic discourse has been successfully made—from a position of strength.

The second type of first contact, unplanned, guarantees the preponderance of forces will be unbalanced based on chance alone—if one side has more forces than the other, it will be because of sheer luck. Unplanned first contacts are dangerous, limited in communication capability, and frustrating in resolution. Each side will attempt to keep as much distance as possible between the other and call in reinforcements as rapidly as possible; the distances each force will keep between the other will likely be as far as they can, possibly even refraining from making contact and retreating from the contact completely. Unplanned first contact is totally unpredictable, which makes it the most dangerous kind. Any military skirmish or accidental damage or loss of life resulting from unplanned first contact will be emotionally charged and very difficult to ignore by both civilizations, especially when attempting to build a brand-new diplomatic relationship.

Assumptions

To make our discussions easier, this book makes a few assumptions regarding the environment and technology available to any interstellar space force. In reality, there are no guarantees.

The Naval Model

As discussed above, any forces serving in space would no doubt be comprised of forces structured in a similar fashion to our navies here on Earth. The two main reasons why this will be so are at once obvious: the environment, and the requirements of spaceborne forces. Travel through space resembles sailing, and conjure up images of vessels gliding effortlessly through the void. Additionally, space vessels may stop and conduct business wherever they see fit—something no airplane in flight can do, as an aircraft remains airborne at the pleasure of its airspeed and generated lift. The nature of space travel means long, often isolated, and unforgiving movement over vast distances, something familiar to terrestrial naval traditions. The cold and lifeless environment outside is to space vessels as the ocean is to submarines. Space is a three-dimensional environment; in this regard it resembles movement through the air, but to the occupant of any space vessel movement in a zero-gravity environment appears to be the same as travel upon a two-dimensional ocean.

Vessels will need to be large to contain the supplies, personnel, and equipment needed for spaceborne mission sets. Ships will also need to carry spare and standby equipment to make repairs if they should encounter difficulties while underway; the unforgiving environment of space will quickly swallow the unprepared and undersupplied for the slightest miscalculation. This is something terrestrial navy vessels already do.

Since spaceborne forces will likely be away from home stations for long voyages given the tremendous distances and relatively low speeds needed for deep space operations, classic naval discipline also lends itself well to spaceborne forces. The need for strong command amidst the vast emptiness of space, the practice of carrying and safeguarding large amounts of provisions for large amounts of personnel over long periods of time, and the already-established rank structure and traditions would mesh well with spaceborne travel. This book will address all spaceborne forces (unless specifically stated) as naval in form.

The State: The Founder, Patron and Explorer of First Resort

Despite all this talk of extraterrestrial competitors, the first foe we will find in space will be us. As we continue to reach for the stars over the next few decades to solve economic and political problems we have here on Earth, those that successfully find what they seek in space will likely be there under a terrestrial state's flag. Even our first permanent outposts and potentially even planetary settlements will likely be state-owned or dominated by a single state. In the end, much like the United Nations or the International Space Station, one nation always pays a greater burden of effort or cost which entitles them to a larger than average share of control or power. Whether this is publicly recognized or not is irrelevant; it is nevertheless true, and governments with the preponderance of control always seem happy to remind others of this fact. States will be the entities marshaling their wealth, information, and manpower to make the first expensive and dangerous moves into the cosmos; there is no realistic alternative.

There are alternatives in the form of civilian space flight—but these are *unrealistic* alternatives. When it comes to first-of-its kind space adventure, the government is the only real possibility. The reasons governments will be the driving force behind space ventures is because of expense, risk, authority, and competition. For instance, even though it may appear civilian space flight is within reach for ultra-rich space tourists, the reality is that civilian spaceflight accomplishes nothing militarily useful, is not space exploration, and does not utilize space resources for the good of a nation. While experimentally interesting, unarmed suborbital out-and-back sorties are militarily meaningless, and a ship full of rich tourists does not a political option make. From a national security standpoint, it only amounts to applause. Rather, just like the West's past ventures from Eurasia to the New World, exploratory and colonial projects will be first made under the auspices of the state. Only a state possesses the resources to adequately fund and absorb potential failures of significant space projects. Corporations did not build and conduct the Apollo program; universities did not fund the Hubble Space Telescope. Only governments possess the requisite moral authority and staying power to blaze the trail. Once blazed, though, it is realistic to expect civilian and corporate copycats and camp followers to appear, just as today's nascent suborbital flights in many ways come on the coattails of NASA's Gemini, Apollo, and Shuttle programs—all of which were funded and risked by the government.

States are the only entities which can remain solvent after the disasters which are sure to occur as we step tentatively into this largely unknown and dangerous environment. Much like accepting risk which comes with war, the state is the only political entity which has the authority to safeguard its and its participants' political independence, make life-and-death policy decisions affecting the whole nation, and secure resources needed by the group for its survival. First-of-its-kind missions into foreign environments have nearly always been funded and resourced by the state and often commanded by state employees. Columbus' expeditions to the New World, early English colonies in North America (and for that matter, colonial expeditions elsewhere), exploratory journeys to Antarctica at the turn of the 20th century, and every space exploratory venture taken to date were all sponsored by the state at considerable cost and risk. First-of-its-kind is always the most expensive and difficult adventure. After all, what do you bring to some place you have never been, nor ever seen? And how much of it? How long and far should you go before turning back? How hard should you push after encountering difficulties? Furthermore, how dangerous are those difficulties? Even in our earliest and most simple space flights, we have already seen minor maintenance gaffes cause tremendous failures and tragic losses of life. What kind of analogous minor detail is waiting to pounce on the first intrepid outer space explorers? What is waiting for us as we venture further? Only the state has proven able to bear these costs, risks, and face these fears.

To prove it, first examine and then discount the next-likely organization which could remotely be capable of first-of-its-kind space ventures: the corporation. There are some who believe that with enough ready cash, just about anything in space can be done. In cyberpunk literature, William Gibson and similar writers prominently paint mega-corporations as non-state rulers of dystopian worlds run by crime syndicates and might-makes-right corporate boards of directors, assisted by endless funds and illicit technology.[34] But any of Gibson's corporations, just like ours, are immediately hamstrung by their most prominent driving need which the government never has to worry about: profit. Even an extraordinarily wealthy corporation could never free itself of this imperative, even if it could someday rival the manpower marshaled by a government. In our universe, unlike Gibson's, corporations lack the legal apparatus to force needed manpower into existence without turning to criminal activity (unlike the state with, for example, conscription). No matter what circumstance or bold mood a corporation's board of directors may find itself in, the corporation, like a government, must still provide whatever resources it needs for its own existence at all times. This means no matter how wealthy it is, a portion of its wealth must always be used maintaining its life support—profit—and its infrastructure here on Earth. These imperatives further fraction its wealth into less-usable chunks available for what we already know to be absurdly expensive space projects. Moreover, in planning its massive project, it must determine how *the space flight itself* will turn a profit and thereby justify itself to the corporation's existence, its shareholders, and the law.[35]

Indeed, when a corporation runs into trouble, to where does it turn for help? The government, of course. When corporations declare bankruptcy and their board members slink away, governments must pick up the pieces and bury the bodies. Where corporations gamble away their fortunes on absurd space projects for which they are unprepared, the government buys their lifeless husk on the cheap and improves the next mission while leveraging its moral authority and primacy. The next recruits sign up with the survivor. Where corporate leaders shuffle away to consulting or lobbying jobs after their expedition fails and their companies dissolve, the government retains its project leads for the next task. Corporations get one shot; governments get infinite attempts. For these reasons, corporations will certainly always follow the government wherever it goes, but can never be counted on to blaze the trail into great unknowns.

Feasible Technology

While science fiction and real scientists across the world are doing their best to test the limits of current technology as it applies to space operations, this book refrains from calling upon technically infeasible or theoretically impossible technology for the sake of focusing on strategic discussions regarding space warfare. While it is true advancements in propulsion, daily existence in space, and weaponry are certainly welcome and probably necessary to make full-fledged space operations possible, for discussing space warfare at the strategic level they are immaterial. Technologies like warp drive, wormhole travel, propulsion at or beyond light speed, automated factories, and other possible but not yet probable topics will be left to science fiction writers and casual space enthusiasts. While there have been recent advancements in artificial intelligence and automated forces to warrant discussions about them, they and other technological concepts like them will remain theoretical throughout this work.

2

Logistical Requirements and Realities

"My logisticians are a humorless lot. They know if my campaign fails, they are the first ones I will slay."

—Alexander the Great

Supplying Space Forces

Perhaps the greatest challenge to any spaceborne force is its maintenance and resupply in deep space. All the great military accomplishments in history—Hannibal crossing the Alps, Napoleon's march across Europe to Moscow, Nimitz's Pacific fleet embarking across the Pacific—were supported by tremendously complicated logistical arms. Even armies which foraged as they went had to carry things which could not be obtained locally: ammunition, artillery, cooking utensils in quantities needed for an army, camp tools, and so on. This means any space force, depending on mission and scale, could become larger than anything humanity has needed ever made in order to provide care and feeding for any human assault forces, officers and crew, spare parts and specialized troops to care for automated forces, and do so over distances—and cruise times—we can barely fathom.

Supplying armed forces is difficult to envision for non-professionals. Military historian Sir Michael Howard notes the modern problem of supply began in earnest in the 18th century, when armies paid for exclusively by monarchs had to confront supply problems for forces which now numbered in the tens of thousands.[1] In particular, Howard notes "the problem of keeping an army some 70,000 strong provided with a continuous flow of food, fodder, and ammunition as it moved through hostile country was the first which the general had to learn to master, and many never got beyond that."[2] Supply is also a measure of speed for any armed force; if the supplies cannot be reunited with a force on the move, that force is usually obliged to stop and wait for them to catch up. This has been an issue for every army in history: Caesar plunged ahead of his supply trains into the feral wilderness of Gaul during his campaigns, with the understanding they would catch up; Napoleon often pressed ahead to give battle with the forces he had, and expected his army corps' individual baggage train to follow the armies at best speed. Even in 1870 Europe during the Franco-Prussian War, a period of relative technological development where both sides had access to railroads, Prussian armies were obliged to wait for their 6,000-strong wagon train which carried engineering equipment and ammunition.[3] Famously, the Wehrmacht's mobility was severely hampered by Allied attacks on fuel and petroleum stocks in the later days of World War II's European campaign.

The numbers and types of supplies which are involved with an armed force of any size are usually nonintuitive beyond the most obvious, like food and ammunition, and in amounts which typically boggle the mind. In reality, the number and types of supplies needed for any armed force stretch from the obvious to the mundane. While the layman might be able to guess approximately how many bullets an army of such-and-such size requires, how many raincoats would it need? How many toothbrushes? What can be prepared on the march, and what can be foraged along the way? These things are impossible to know without professional training.

Luckily, there is great precedent for professional supply services, which will certainly be necessary for future space armed operations. The U.S. Navy first commissioned its Supply Corps, an entire branch of the Navy dedicated only to supplying its fielded forces, in 1795.[4] The Supply Corps has served over two hundred years continuously supplying naval forces with everything from uniforms to entire vessels, and naval forces by their nature are constantly deployed, which has presented a special challenge to Supply Corps operations. Even in times of lean budgets, like the dawn of the 19th century and America's battle with the pirates of the Barbary Coast, the Supply Corps did its best to provide sailors with their meager twice-a-day meals of dried meat, raisins, hard biscuits, and grog.[5] Beyond this, the Supply Corps was also responsible for paying all bills and procuring the necessary means to make war, such as gunpowder and saltpeter.[6] One vessel, the USS *Brooklyn* at sea in 1898, illuminates a typically fleet supply example of the day. The *Brooklyn*'s complement of 33 officers, 20 chief petty officers, and 427 men received an average monthly pay of $20,000, and issued on average 2,000 pounds of soap, 100 cap ribbons, 50 pairs of shoes, 50 sets of underwear, 25 pairs of trousers, 25 shirts per month, and also took on 500 pounds of tobacco purchased and shared by the crew.[7]

As time went on, wars conducted by the United States further and further afield challenged the naval logistics system. During the Gulf War in 1991, the Supply Corps found itself overextended in what had clearly become a logistics-heavy war that depended heavily on sealift to supply its aggressive air campaign and transport U.S. ground forces to the theater.[8] The 10,000 miles and 12 time zones that separated the U.S. east coast and Iraq stressed communications between deployed forces and the naval supply stations, and it became increasingly harder to determine what was needed in a timely manner.[9] Extrapolated for bigger vessels, more crew, better technology, and longer voyages in space, it is clear supply will be a daunting issue.

Then as now, Congress' "proclivity to micromanage military budgets" did not sit well with the U.S. Navy, nor with any other military force, and naval supply facilities like wharves and ports were easy targets for congressmen opposed to standing armies and navies.[10] One particular congressman, Hon. John Randolph of Virginia, happily launched a "fishing expedition" in 1809 to search far and wide for naval misappropriation and graft.[11] While he seemed ostensibly concerned about mismanagement, his real aim was to embarrass President James Madison and his cabinet.[12] The political infighting, which eventually led to re-commissioning the Navy's pursers in 1812, was an unnecessary distraction as war loomed on the horizon with Great Britain, finally breaking out later that year. Political maneuvering will never disappear, and it is a lesson we must take with us as we move closer to procuring standing space fleets and forces.

Beyond basic sustenance, fuel is also a major supply issue for space forces. As the U.S. Navy gradually built more stream-driven vessels at the turn of the 20th century, coal consumption grew rapidly and was an instant supply problem. When coal was new, the Navy depended on manual labor to resupply the fleet. A typical collier vessel of the era required 100 men to bag coal inside the cargo hold of a coaling vessel with little to no protective equipment.[13] Once bagged, the coal could then be transported to another vessel, which normally took about 10 hours to deliver 1200 tons.[14] The coal itself had to be procured domestically, bought internationally, and then itself transported overland or by coal-burning vessels to U.S. coaling stations, making coal transport and storage expensive and a significant part of the Navy's budget.[15] This practice clearly needed to change if the U.S. Navy were to be less dependent on frequent coaling and to become more globally maneuverable. Outside of the east and west coasts, the Navy possessed coaling stations only at Honolulu and the Philippines, further raising expense by often requiring naval vessels to pay to use other nations' coaling stations.[16]

While spacefaring vessels will not burn coal, the example of the early

U.S. fleet adjusting to global operations is a useful analogue for future space transportation. Unlike terrestrial navies, though, space fleets which run out of fuel will become "dead in the water" in more ways than one. Since fuel in whatever form will be necessary to run the life support and propulsion systems on spacegoing vessels, to run out of fuel means to also run out of life. The establishment and strategic placement of interstellar "coaling stations," whatever they look like, will be an instrumentally important logistical concern in space.

Forces: To Automate or Not to Automate?

What would a space force look like? Science fiction has done its best to mimic current naval forces in appearance, command structure, and even tradition. But given what we know—and don't know—about space and the stresses it places on personnel and equipment, what would our craft and space forces end up looking like?

Compared to terrestrial forces, space forces will be particularly prone to logistical shortfalls. Much like today's naval forces, spaceborne vessels must carry with them everything they could possibly need or carry the means to produce what they need. Moreover, unlike terrestrial forces, the potential for foraging enroute is non-existent; space in its hostility and wilderness is not prepared to feed and care for a human host of any size or composition. This mean that the more the military requirements are examined in light of traditional military calculations, the greater the lure of—and perhaps need for—automation becomes. After all, machines don't eat; unlike certain science fiction works which advocate forms of chemical hibernation to allow humans to travel unconsciously with minimal metabolic processes over long distances, the rapid pace of our current era's artificial intelligence (AI), cyber technology, automation, and computer technology advancement may preclude any need for long-distance human travel other than peacetime activities. Indeed, we may find ourselves not needing to send large invasions or expensive fleet forces into the stars; not when robots can do the dirty work for us.

Thus, logisticians and military planners have a choice: automate, or do not. Ideally, choices about force composition will be based on strategic objectives and what the force in question has set out to do. The most consequential result of force composition is its tactical results; in other words, the shape of the tool chosen has a direct impact on the job it can do. Hammers cannot tighten screws very well, and screwdrivers make poor hammers. Similarly, bombers make poor cargo aircraft, and airlifters find bombing a challenge.

Nevertheless, both could perform this reverse task in some fashion, though it would be inefficient and ineffective. This means the space strategist must begin with the political objectives, then work towards making the best force composition possible based on what they have available.

Regardless of what we decide to do, it is clear manpower in the form of actual human participation in military space activities may be the most available resources we have for some time. After all, even if AI advances to the point where it can be safely relied upon for military activity, robotics must subsequently catch up; and any reliance on an automated force doesn't begin to address the problems associated with machines being absent from their maintainers for a long period of time. Further, it would behoove us to initially man any spacegoing vessel with a larger than minimal crew complement. The reasons are at once clear: given the doubt and uncertainty surrounding initial deep space operations, additional crew will be necessary not only to accomplish adequately a ship's primary duties while away from port, but also to be ready to absorb additional casualties while far away from home and completely surrounded by a hostile environment on the best of days.

Space forces, then, will likely take one of the following general three forms.

The Terrestrial Analogue

This force would be of a form most familiar to us: space vessels and armies completely manned by human officers and crew and supplied in a traditional manner, that is, carrying what they need to arm and care for those aboard, and obliged to rendezvous with regular supply and fuel ships. The crews would function with as much automation as governments deem necessary, but by and large would perform all of the ship's critical functions. Maintenance, supply, medical care, repair, battle, life support, research, leadership, and travel would completely be the captain's responsibility. Naturally, a vessel or fleet of vessels such as this would have a great deal of autonomy, as is reflected in naval tradition. Namely, this means command and control of these forces begins with a general order given to the fleet admiral (or captain for solo ships), with interpretive authority then flowing to the commanding officer with little to no higher echelon of command oversight.

Pros:

- **Familiarity.** Humankind is well acquainted with this model of naval operations and it fits well with human social and leadership structures of any culture.
- **Maximum operational flexibility.** This model ensures commanders in the field possess the greatest operational flexibility to decide the proper course

of action for their crews and their vessels. The human element is not only present for a combat situation, but is necessary to effectively conclude one.

• **Maximizes morale.** In deep space, large crews of like-minded personnel (e.g., highly-motivated spacefaring wannabes) would provide a tightly-knit company, as opposed to personnel forced to spend their days with automatons or drones. The lure of space adventure alone could completely satisfy recruitment requirements.

• **Limited investment.** While it may not seem true at first, a completely-human force could be best tailored for an individual nations' budget and would be cheaper compared to an all-automated force. With resources flowing from a centrally-administered Space Fleet, salaries and support can be adjusted similar to contemporary military services.

• **Maximizes ingenuity and adds to survivability.** No one can solve human-centered problems of survival and adaptation quite like humans. It is very clear our species is exceptionally adaptable, and this will no doubt carry on into deep space, though the benefits of these talents are not immediately clear as of this writing. Regardless, any human force can be expected to do its best to survive in harsh environments and save its vessel—or die trying.

• **Rapidly adjustable in size.** All-human forces and vessels can be quickly manned, unmanned, adjusted, or shuttled about due to our historically vast experience in manning and transferring terrestrial forces. This makes terrestrial analogues mobile and agile, but as always potentially prone to delay, mismanagement, and confusion.

• **Experienced forces act as success multipliers.** The longer a crew or force remains in service, the better they get; the more varied and experienced their careers, the better served a state's space power will be. Naval forces and human crews with more time served in space are better and more effective at battle. Further, scientific discoveries will only continue to blossom from experienced science crews who spend more time in space—another benefit from human adaptability.

Cons:

• **Catastrophe is extremely costly.** Space is not a welcome environment for things to go wrong. Shipwreck is likely fatal. Given the dearth of survivable worlds with breathable atmospheres and the time necessary for rescue to arrive, abandoning a vessel and waiting for rescue is probably not realistic. Solitary vessels are especially vulnerable to space anomalies, equipment malfunctions, or even hostile encounters with superior spacefaring forces (terrestrial rivals or otherwise).

• **Inefficient compared to automated forces.** Much like modern day remote piloted military equipment, automation eliminates on-site human factors and can generally operate longer than manned assets. A human force in space would require everything an ocean-going terrestrial navy would require, and robots don't have to eat. Human crews, even with multiple shifts, need at least one third of their time to sleep, and nearly another third to dine, medically care for themselves, groom, and train. Worst case, that leaves only a third of their total time onboard a military vessel or in a military organization to carry out their duties; this can be stretched in a contingency, but the vast distances and time-consuming nature of space travel will require balance more towards life than work. A smart captain won't push his crew too far when they are several million miles from help.

• **Frequent resupply and transfers limit a force's range.** This is a well-worn lesson taught to us by terrestrial naval forces. In fact, in the early days of deep space travel it will likely be the logistical supply lines, and not a ship's engines or fuel, which limit any spacegoing force the greatest. Like modern forces, any regular spacegoing military force will also require regular personnel transfers: to evacuate wounded, exchange personnel for career progression, and to change commanders and officers.

• **Limited by human capacity, rest, and so on.** As mentioned above, any human-only force's efficiency will likely be limited to approximately a third of its total available time. The potential monotony of a spaceborne cruise harkens back to the early days of seagoing sailing vessels; cramped quarters with little in the way of things to do, barring anything one brings on board themselves. At some point during any voyage the initial thrill of deep space travel will subside, leaving routine and discipline as two rather blunt mainstays of the trip. In any case, time spent onboard ship during such voyages has a limit; we just don't know what it is yet.

• **Likely unsuitable for conscription.** While "Shanghaiing" men to serve aboard sailing vessels was common in earlier times, the nature of deep space voyages does not lend itself to forced service. Sailors against their will onboard spacegoing vessels are simply too dangerous to keep around. Due to the distances of travel, time spent away from Earth and home will most successfully be guaranteed via voluntary contracts instead of forced service. Moreover, voluntary crews are one of the only reliable ways to avoid catastrophic disciplinary incidents, such as crew mutinies or fractious physical infighting between the crew.

The Hybrid Force

Undoubtedly, any space force will require some kind of critical component to run by computerized enhancement or automation to meet the deadly standards required in space. Since the earliest days of spaceflight, onboard computers have been required to perform some functions essential to a successful mission. Most famously, the early U.S. Mercury Program missions were originally designed for entirely automated flight, with engineers discounting pilot involvement and derogatorily referring to the competitively-selected and highly-capable U.S. Navy and U.S. Air Force Gemini pilots as "spam in a can."[17]

Since those days, computers and automation have taken greater and greater roles in spaceflight missions. However, NASA and other nations' space agencies have been keen to recognize the importance of "the human element" during space missions. After all, it was Alan Shepard who needed to modify the pitch angle of his Mercury capsule to maximize his craft's reentry survivability after his ship's heat shield suffered damage, arguably saving the U.S. space program.[18] The ground-based engineers and crew of Apollo 13, working together, solved the critical problems of returning the star-crossed crew and vessel safely home.[19] How many near-catastrophes have been averted by alert crews, with capable onboard computers and engineer support, since then? Only NASA knows for sure.

Pros:

- **Less personnel required for a full crew compliment.** With combination automated and manned crews, clearly a portion of the personnel required to carry out a mission will be replaced by machines or systems which do not need rest and care. This could enhance the chances of success for a spaceborne force.
- **Less support needed than completely-human crews.** Machines and systems, after all, don't need to eat, don't get sick, rarely complain, don't sleep, and have no need for recreation.
- **Provides a secondary catastrophic point of failure.** During colossal failures resulting in either the incapacitation of the crew *or* the loss of onboard systems, the surviving half—either the crew or the automated functions—could be used to return the vessel safely to a friendly port and rescue at least part of the crew or mission. Both man and machine are present in sufficient enough concentration to act on their own if the other becomes debilitated.
- **Crews can retain their desired naval structure, tradition, and preferred operational procedures.** When only part of the vessel or force is automated, there is nothing stopping the remaining human element to retain what is important to them: traditions, operational designs, and general shipboard routines. The automated portion of the force works *with* and *for* a human compliment, not above or against them.
- **Very advanced AI could aid mission completion and combat performance.** It is conceivable that sometime in the near future AI will get so good

that it could be able to demonstrate real tactical results during an engagement on its own. While AI has already proven it can beat human players and other computers at chess and other "tactical" games, the reality remains that actual operations and combat do not possess what those games would call "rules." To be sure, a battle often functions ostensibly under what's termed "rules of engagement" in military jargon, but these are not rules which govern the physical progress of a battle, nor do they direct a combat leader about how they should go about executing a mission. They are simply a list of proscriptions, of "do's and don'ts" which lay out a situation rather than guide you through it. In this regard, any action taken during combat which is not proscribed by these "rules of engagement" is fair game, and requires a creative mind to discover and execute. AI has not yet shown this capability, but it could happen one day, and could give an on-scene commander in deep space an electronic tactician to aid his or her decision-making.

Cons:

- **Eliminates some of the "human element."** From Hannibal to Napoleon, from Nelson to Nimitz, if it's one thing the military understands it's that there is no replacing the *je ne sais quoi* which plucky commanders in forces of all kinds have used to save themselves and their troops from certain death, suddenly outmaneuver a superior force, or execute a strategic campaign so startling in its success that it changed history. Manning any vessels with less numbers of well-trained officers and crew will leave less of the "human element" available to make life-or-death decisions and execute tactical and strategic breakthroughs.
- **Malfunctions require more expertise to repair.** Readers who have worked in military operations have certainly experienced the following situation. A particular component, aircraft system, or computer program, has broken down on a spacegoing vessel. This part, system, or program was installed by the XYZ Corporation, loyal member of the military-industrial complex, who had contracted with the military some years ago with an excellent pitch by a smooth-talking executive. The military, hoping to acquire a capability which the part/system/program provides, now contractually relies upon fielded employees of the XYZ Corporation to provide repairs or support for the part/system/program. The item, of course, is intellectual property of the XYZ Corporation and could never be expected to be handled well by serving military mechanics/maintenance/computer personnel. The trouble is, to save money and win the bid, the XYZ Corp. keeps only a smattering of sub-contracted repair personnel on hand, and only a small number of these are experts qualified to repair the part/system/program. To make matters

worse, these experts are not combat-rated nor contractually authorized to enter combat zones to repair the part/system/program, which is just where your aircraft/computer/ship happened to be stuck when the damn thing broke. Since vessels' crews have been replaced with more automation, uniformed personnel capable of finding a workaround or solving the issue are no longer stationed with the aircraft/computer/ship. Deep space forces would be well served to heed this modern lesson: carry whom you need in case your shiny new industry-produced component decides to kill itself while several thousand light years from home. Reduction in human crews could make this possibility more likely.

• **Malfunctions carry an increased risk of mission-ending failure.** With less qualified crew on board, that means the risk of any of these crew being unavailable when they are needed due to sickness, injury, or even when crew rest cycles change is higher. If an emergency occurs, the ability to get qualified personnel to repair a malfunctioning unmanned or automated critical component could mean the difference between life and death. In general, when it comes to installing automation on spacegoing vessels, it is unwise to create such components or systems which have a single point of failure that is at the same time automated. This means to the maximum extent possible space forces should have manual or manpowered backups to critical systems, or the means to take systems offline and replace them with older but more reliable methods. To have a vessel or vessels which are able to limp back to a support area is much safer than one which could be dead in the water due to a single system failure. In space, of course, there are no currents to ride should the masts collapse; no sea in which to fish; no islands on which to forage or trade for supplies; and for all practical purposes no help which is not weeks or months away.

• **Onboard priorities could force crewmembers to sacrifice themselves for the good of the automation.** It is safe to say whichever individual ships' functions and systems are given to full automation, some will no doubt be life support, propulsion, or defense functions. These three functions are of primary concern to the continual existence of the vessel as a whole; without them, the ship itself and all aboard would perish. This means that given a situation or emergency which is bad enough, the crew may come to harm or even be called upon to spend some of their lives to repair or save a critical onboard component which keeps one of the aforementioned three critical areas functioning. It is not difficult to imagine, for example, an engineering crew being required to focus repair efforts on a heavily damaged reactor or propulsion unit, all the while suffering fatal radiation exposure. A man-machine contract of this sort already exists in today's terrestrial navies, but the difference in space will be the degree to which automation will be relied upon to carry out critical functions, which could lead to a sense of urgency requiring a firm commitment of manpower to a problem which, if left unaddressed, could leave the vessel in question adrift hundreds of light-years from help.

The Totally Automated Force

Why go yourself, when you can boldly go where no one has gone before from the comfort of your armchair? The draws of automation during space flight were immediately clear from the start of manned space missions. Given the long missions, distance, and a cold and lifeless deep-space environment which saps human vitality through its long stretches of inactivity, it was immediately clear certain spacecraft systems would require automated components with a reliable shelf life in order to travel in the cosmos' most austere environments.

But could an entire force be completely automated? There's no reason to believe they could not; it is easy to imagine fleets of robotically controlled troop carriers and combat vessels, packed to the gills with machines and weaponry, needing no life support systems nor atmosphere of any kind, with plenty of room for redundant components and maximum firepower. While the most eager explorers among us would cringe at the possibility of an interstellar voyage which would leave them behind, is there military merit in the completely automating the force? If history is our teacher, it is profoundly dangerous to completely remove the human element from any war machine, from the lowliest email server to the largest aircraft carrier. Nevertheless, the idea of an android fighting force deserves its own examination.

Pros:

- **Completely removes the need for life support and organic supply.** Machines, as stated above, need no food, water, rest, entertainment, or atmosphere. A vessel's cargo and composition can be maximized to include as many machines and weaponry as possible, and they can be crammed into vessels with no regard to comfort.
- **Vessels can theoretically operate with minimal rest or refit.** If properly crewed with repair or upgrade mechanisms, a fully automated vessel could simply be sent blueprints with software-based instructions to take care of its own repairs and refits. Vessels would only need to return to port to resupply any building materials or armaments. Alternatively, automated vessels could never return to port but be met by other automated repair or upgrade vessels while enroute to take care of any overhaul needs.

- **Preparation time is limited only by the speed of industry.** Once an automated force is built, little prevents it from entering active service immediately. Like a toaster out of the box, once it is plugged in and situated properly on the counter, it is ready to go.
- **Machine-only forces cost considerably less than manned ones, if truly unmanned.** As of this writing, "unmanned" forces are rarely that; a legion of communications specialists, satellite mechanics, unmanned pilots, sensor operators, and aircraft maintenance personnel are required to fit and fly any unmanned force. Conceivably, if the manned portion of any unmanned enterprise can be minimized, the force becomes cheaper than manned forces, even when including the initial investment for construction.
- **Can traverse potentially dangerous areas which manned spacecraft could not.** Unmanned machines have much less to fear from radiation, solar coronas, high gravity, noxious gases, and other harmful stellar phenomena than do manned crews. The effects of these phenomena on machines may be unknown at first, but once properly understood they will likely present no obstacle to robotic or automated forces.
- **Advances in AI could mean automated space forces capable of learning from their mistakes and experiences.** If machines are truly capable of learning in the same way humans are, in theory automated tacticians could best any human tactician at deep space command, provided they survive enough encounters to get there. Debate on this issue is fierce, but if machine learning is possible it could apply to automated deep space forces.
- **No human life lost in case of disaster.** Clearly, a nation fielding an automated force has much less to lose politically and less economically if they don't have to worry about getting a great number of their citizenry or constituents killed. Indeed, totally automated forces have no front-line force recruitment issues either.
- **Reliable onboard systems decrease chances of failure.** Note the statement; more automation does not *increase* chances for success, it only decreases the chances for failure. This is because onboard automated systems' reliability is only increased through upgrade or redundancy. Contrast this with putting a better educated or more experienced human than the crew previously had on board a ship. The upgraded piece of equipment may be able to do more tasks faster, but a better human can function beyond his programming parameters.

Cons:

- **The human element is completely removed.** As stated above, there is an ineffable quality to having humans with human brains—the ultimate computer—nearby and ready to react and adapt to emergency situations. A totally automated force will have no human eyes to gaze outside the portholes; no Captain Sullenberger to execute a brilliant emergency landing in the Hudson River; no Nelson to out-think and surprise an enemy; no Olds to dream up a new tactic just when the enemy least expects it. The automated forces would possess no benefit of experience; only programs created, at some point, by a flawed human programmer.

- **Entirely automated forces are at the mercy of their weakest component's reliability.** An old military adage states that an army is only as fast as its slowest soldier. Similarly, a computer is only as fast as its weakest or slowest component; what good is 600TB of RAM if it's shackled to a 200 megahertz processor? In this fashion any unmanned force will function as fast as its least-reliable component, which will likely require constant vigilance by human handlers.
- **Repair and refit, if needed, will be difficult to diagnose and respond to.** This is especially true if the diagnosis equipment itself is faulty, if the force has traveled far beyond a frontier or repair area, and if any trouble includes faulty communication protocols or equipment. Any program or ability to diagnose an unmanned force from afar will have to be absolutely ironclad in its design in order to minimize doubt in diagnosing the problem, and such diagnoses must be absolutely sure before dispatching repair crews (which suck up additional military escorts). As these maintainers could potentially be light-years away, it would be a shame if they brought the wrong parts to a stranded and unresponsive automated vessel.
- **Totally automated forces are completely reliant on command and control signals from a distance, which creates a critical vulnerability.** No matter what form an unmanned communications hub takes—a structure, an organization, a group of networked systems, or other vessels whose sole mission is command and control—the communications hub itself will be vulnerable to the same problems any equipment have been throughout history. They also represent the primary connection between the unmanned assets and their human masters, and therefore become a primary target of any adversary. Removing command and control is one of the fastest ways to turn an unmanned space force into man-made asteroids. If a state is foolish enough to deploy unmanned space forces *without* any kind of human-controlled failsafe or command and control capabilities, whatever logic the unmanned force uses to generate its own command and control decisions would be the next best target.
- **Incapable of diplomatic discourse if encountering other sentient beings.** Robots make poor bedfellows, and tend to lack charisma.

• **Complete conquest of a territory is unlikely due to local control.** Should an unmanned force be ordered to conquer a system or other similarly interesting military target, beyond simple destruction and occupation an unmanned force will likely be useless and will require a human presence to completely finalize an occupation. A force which cannot negotiate with the enemy, cannot look beyond material and tactical concerns, and which cannot creatively think will not long be effective during occupation duty.

• **Most contingencies must be thought out beforehand for programming purposes.** Barring any revolutions in AI, any unmanned force will have to come equipped with a formidable amount of calculation algorithms, software, and reactive authority to deal with a contingency in approximately the same manner a manned crew could. In short, if it doesn't exist in programming and if the computer hasn't been taught to deal with it, then an unmanned force is likely to fail in its reaction to an unprogrammed incident.

Regardless of what we ultimately decide to do, early spacecraft and spaceborne forces will doubtless depend heavily on highly-trained manpower supplemented by the most advanced technology we can muster. After a few decades of experience, where we go next will depend largely on cultural influences, technological advancement, and a somewhat less discussed but stubbornly persistent teacher: lessons learned from major disasters and accidents.

The Attrition of Distance

Distance creates trouble before campaigns even begin. From our experiences here on Earth, we know that any force on the march or which sails the sea, no matter its composition or type, begins to suffer attrition of some kind before it even arrives at its destination. This is especially true if the territory it crosses is hostile. This attrition can be as small as minor equipment malfunctions, or as major as losses rendering a combat force incapable of fighting. During transit, forces are generally more vulnerable to attack or interception than when deployed in a combat-ready stance, which means the longer a force is traveling the more exposed it is to interdiction, hazardous phenomenon like weather, and equipment failure. In general, neither distance nor time are neutral: the longer a force is in the field and the farther from a support source it is, the more there is that could go wrong. Attrition manifests itself in many ways: soldiers on long marches get sick or desert; ships break down far from port; aircraft encounter ferocious weather affecting their flight plans and arrival times.

Attrition is the unavoidable loss of military forces due to sickness, injury, accidental death, desertion, equipment malfunction, weather, or any other casualty-causing factor which does not include battle. Specifically, attrition of distance is a concept long known to military logisticians that refers to non-combat casualties caused by what happens between the beginning of the campaign until forces are disbanded. The greater the distance, the longer a force has to travel, the more time it spends exposed and unsupplied. A shorthand definition of "attrition of distance" would read: "the inability to avoid casualties the farther and longer you go."

What would attrition due to distance look like in space? Desertion is nigh-on impossible; it is doubtful anyone could somehow sneak away from a fleet or force while underway due to tight controls on methods to exit a vessel, and in no small part due to the unfriendly environment of space. Sickness is certainly possible, but being confined on vessels, it would likely be of terrestrial origin and therefore treatable. The long cruising times needed to reach an objective via space flight, combined with intense health screenings prior to debarkation, make sickness only a minor concern while underway on a space campaign. While some argue that an extraterrestrial virus or bacteria would be unable to sicken a human, it is not immediately clear what impact encountering an extraterrestrial organism would have on a force in motion.

Classic logistical and disciplinary problems are more likely than any other factor to cause attrition during a spaceborne mission. Insufficient food and water, waste and spoilage during the journey, poor organization, illicit drugs, mutiny, mechanical decay and failure, and accidents are all distinct possibilities.

Cruises, especially during our initial exploratory era, will surely require long-term service with less than ideal entertainment facilities. The classic way to deal with the problem of boredom is to keep the crew occupied with their duties as much as possible, even to the point of inventing work for them to do, which is preferable than idleness aboard a vessel thousands of miles from an atmosphere.

The lifeless vacuum of space will in many ways act as a cocoon that surrounds spaceborne forces and prevents the influx of undesirable factors, while deterring any deserter from braving the cold darkness to flee into a lifeless void. These facts will help prevent classic force attrition, but also remind us that reliable spaceborne supply systems and precautions must be established before spaceborne forces begin extended operations. There are likely also many unknown deleterious attritive effects waiting for spaceborne forces to unfortunately discover during our first long cruises. For instance, one unique form of attrition we can expect to affect spaceborne manned forces is that of minor forms of mental instability, even dementia, due to the stresses imposed on a human psyche during the protracted time required to traverse the vast distances of space. As astronauts continue to test long-term livability in space during extended sojourns above Earth, we will learn more about long-term effects of living in space which we can apply to future force calculations.

Supplying a Planetary Invasion Army

It is difficult enough to supply a spaceborne naval force underway and far from home. While some films and science-fiction novels make it look easy, or write it off altogether, the plans needed to execute a planetary invasion would dwarf any other plans any military planners have ever conceived. The logistical piece alone would likely be one of the most complicated plans the human race had ever devised, with timetables that would make the Schlieffen Plan look like a bus schedule.

A planetary invasion army represents the "worst case" logistical situation, and is chosen here as an exercise to examine logistical concerns via a specific and challenging example (planetary invasion will be discussed from a tactical perspective in Chapter 5). In other words, going through the needed thought to logistically prepare a spacefaring planetary invasion army is a good exercise to begin thinking about the challenges facing any spacefaring force in general, no matter the target.

For our calculations, let's decide on a above-described Hybrid Force for our invasion army. After that, asking some initial questions help us cage the problem.

First, who would comprise the human parts of the invasion force? Regular army volunteers, or conscripts? Would volunteers even understand what they would be getting into, or be prepared for interplanetary warfare? After that, what would be the percentage of robotic soldiers in an invasion force? There are few terrestrial analogues that could conceivably help prepare a planetary invasion force prior to operations.

Second, how many total soldiers, human or robotic, are required? Planetary invasion would, presumably, require a planetary-sized army. In the case that the target planet is the approximate size of Earth, an invasion army could require soldiers in the billions to make the assaults, absorb casualties, garrison the planet, put down resistance, and dampen any political mess after the conquest is completed.[20] Each continent inhabited by enemy forces would need to be occupied and policed; each continent *without* habitation would also probably need to be occupied in order to deny the enemy places to establish bases in the wilderness. The seas and skies of the target planet, if traversable, will require force superiority to make them safe for our assault troops.[21] In fact, it is possible that planetary invasions could require an Earth government to empty entire planetary settlements of soldier-aged citizens to satisfy the required number of soldiers needed to fill out a planetary invasion force. It is doubtful Earth's population alone can support the production of food and war materiel required for such a force, and at the same time also provide the needed troops to invade an entirely separate planet. This means policymakers and war planners who want such a capability must argue for and support human settlements elsewhere in our galaxy, which will provide answers to the mathematical realities facing logisticians planning a planetary invasion.

With all these people and all this stuff to bring along, space transport becomes a significant challenge and the size and number of vessels required balloons quickly. As the size of the support force needed increases, the ability to conceal the invasion force also diminishes; all the while, each and every vessel supporting an invasion is susceptible to the attrition of distance with every minute underway. Further, millions of soldiers stuck on spacefaring vessels constitute an obstacle to the crew; soldiers awaiting invasion will likely possess few duties while underway other than physical training, studying the assault plan, and staying focused on their grim duty. There could conceivably be millions of soldiers at one time on many different vessels standing relatively idle, which only adds to a captain's problems as some soldiers will no doubt fill this idle time with activities contrary to good order and discipline (as idle troops are often wont to do).

One potential solution is virtual reality (VR) technology. As VR technology continues to develop, VR provides an excellent solution to both long cruise boredom and real training concerns. Moreover, VR can be used both before dispatching the force and while underway. The target planet and planned battle areas could resemble environments found on Earth, but there is no guarantee and little likelihood the target planet will conform to humanity's preferred climates, thus presenting challenges to realistic training on Earth before the assault force departs. If the target planet is not a planet at all, but an enemy outpost or base installed on another object like a planetoid or moon with a hostile environment, that probably means the enemy is also fighting against the same tactical restrictions our forces are. Indeed, this can be a tactical advantage in the right circumstances. In any case, the time spent during the long voyage to the target can be filled with VR-powered training events for the assault force. The more training the troops undergo, the better; and such training has the added bonuses of keeping the potentially millions or billions of troops occupied.

Logistical Impacts on Fighting in Hostile Environments

As discussed briefly above, there is no guarantee our chosen battlefield will possess the minimum requirements to support human life, which will oblige any human assault force to bring their own life-support equipment or the means to create an environment suitable to human life. Moreover, even if a life-capable target planet supports respiratory organisms which breathe a nitrogen-oxygen atmosphere, that does not mean we would want to automatically consider it safe to do so. The effects of breathing in a foreign atmosphere which contains bacteria, organisms, organic compounds, and a whole host of completely unknown and possibly dangerous chemical substances may preclude our human forces from breathing on their own until we are absolutely sure the atmosphere is safe.

But there is more to hostile environments than breathing; even the choice and shape of a battlefield may be limited. In many solar systems, there are little more than lifeless balls of gas or rocks silently marching around their sun or suns. The former is almost completely useless to us for force deployment, but can provide tactical deception as we will discuss later. The latter may be the only place in a system where troops can be set down on something resembling solid ground, but remain hostile to life in any case. Indeed, it is much more tactically easy to assault small and hostile outposts such as these, which require less troops and time to seize or destroy, than it would be to seize an entire enemy planet. The more life a planet supports, the harder it will be to take. The hardest target, then, would be a non-human enemy's homeworld which is both hostile to human life and fully populated. Assaulting this kind of target carries with it special concerns.

Atmospheric Concerns

Perhaps the most obvious concern when planning a planetary invasion is there may be a good chance the target planet's atmosphere itself will fight the invading force. The logistical impacts on fighting in a poisonous atmosphere complicate operations. First, any facility, piece of equipment, or personnel placed on a planet with a toxic, un-breathable, or harmful atmosphere is instantly at risk, at all times and without respite, for decompression, decay, damage, and total loss of life in the case of catastrophic support system failure. This makes any terrestrial force support structure or equipment a superb target for the enemy. These conditions make human forces increasingly susceptible to enemy sabotage and grants the enemy increased flexibility, since the enemy would presumably operate with little restriction anywhere on its own planet. This also presumes the enemy is at home or in an atmosphere hospitable to them, and would not apply to a battlefield equally inhospitable to all fighting forces (the surface of a planetoid with no atmosphere, a moon, or an exposed space platform, for example).

Mitigating this logistical problem will be extremely difficult, and strongly recommends the idea that planetary invasions should be organized, supplied, and commanded from orbiting friendly vessels to the maximum extent possible (this concept will be discussed in greater detail in Chapter 5). Nevertheless, it is troops which take and hold territory, and the need to operate in a hostile atmosphere requires creative thinking. One solution is rotating out-and-back patrols from the aforementioned orbiting facilities, similar to combat patrols performed by U.S. Army troops during the Iraqi pacification of 2003–2012. If the atmosphere is a nitrogen-oxygen mixture but not in the proportions we require, a simple additional breathing device could be all that would be necessary, especially if that device were able to re-mix and sanitize the air entering the apparatus itself. More toxic environments could require full pressure suits, which could be extremely risky during combat if punctured or damaged. One creative solution provided by science fiction is the fully-enclosed powered armor suit, enabling a soldier to survive hostile or vacuum environments while also enhancing his physical fighting abilities through mechanical augmentation.[22]

Finally, not just the atmosphere but the local weather and differing radiation from the target planet's sun or suns will also, more likely than not, adversely affect an invasion force. Both of these problems are constantly present, during work and rest cycles, while eating, fighting, and soldiering, and being locked in a pressure suit or safe facility after a long journey to arrive at the target planet will add to the discomfort and misery of the invasion force, chipping away slowly at morale.

Gravity

Depending on its location within its solar system, the planet in question may have significantly different gravity levels than our terrestrial troops would be prepared to endure. Some would make the case that since most Earth-like planets we've discovered thus far would be the most likely candidates for conquest, any future rivals' Earth-like planet would therefore be approximately the same distance from its sun(s) that Earth is, which would lead to approximately the same levels of sunlight, water, and at first glance, gravity (as gravity is a function of distance from larger bodies).

But this assumption would be a mistake. Gravity is not only a function of the distance of a body from other bodies, but also a function of its mass. As any schoolchild knows, one weighs less on the moon because the moon is less massive than Earth, and thus objects closer to the moon (i.e. those standing on it) are affected less by Earth's higher gravity than those standing on Earth.

Gravity is a function of the distance of an object from that of other objects, as expressed in the formula

$$F = G\frac{m_1 m_2}{r^2}, \qquad G = 6.67\times10^{-11}$$

which is essentially a complicated way of saying the force of gravity exerted on any person or object is directly related to the mass (m_1 and m_2) of that object, and its distance to other objects of larger or smaller gravitational force. This means when it comes to the gravitational effect on our forces while campaigning on distant worlds, it depends completely on the planet in question. Fighting other humans in challenging gravitational environments would be an even playing field, but biologically there is no conceivable limit to the gravitational conditions under which a potential non-human sentient rival could evolve. To us, a sentient rival's world could span a range between extraordinarily light or bone-crushingly strong gravity. To them, it would be home.

This means any invasion force targeting a planet with substantially different gravity levels than Earth would likely have to be mechanically augmented, either through personal means (e.g., some kind of powered suit or machine assistance), or avoided altogether by using automated troops specifically engineered for the gravitational environment. Robotic forces have the advantage of being able to be manufactured with the gravitational effects of the target planet in mind. Depending on how close the target planet's gravity is to Earth's, the gravitational difference could conceivably be trained into human invasion forces through muscular exercises and VR training. While this would take months to accomplish, such training would be an ideal pastime during the long voyage enroute to the target planet. Any difference in gravity between what the invasion force is used to, even in the slightest degree, could spell the difference between victory and defeat once battle begins. Absolute familiarization to the target planet—making it the force's world—is vital to any chance at victory.

Seas, Skies and Lack Thereof

When asking a casual observer what a planet would look like which humankind may someday find worthy of invasion, the first thought which pops into anyone's head is "earthlike." After all, what would be the point of invading a planet with no intrinsic value to humankind?

A better answer, of course, is there are many good reasons to invade planets or planetoids which may not be earthlike or even habitable. In any military campaign, objectives are seized which generate the greatest strategic value; picture a traditional terrestrial army seizing a high precipice or mountaintop simply for the altitude value alone, which acts as an exceptional observation point and artillery station. Terrain which is completely valueless otherwise has been seized by many a force in order to dish out the greatest punishment to the enemy via its tactical superiority (which therefore has an advantageous strategic effect). Indeed, for several hundred years utilizing terrain which was not particularly useful nor habitable on Earth, and which possesses no great value in peacetime, was the main prerogative of the operational plans of many armies. One good example from the Russo-Japanese War of the early 20th century describes the Japanese Imperial Army absorbing an abhorrent number of casualties to seize the high ground surrounding Port Arthur, a Russian fleet and military stronghold.[23] Once it was grabbed, the high ground acted as a critical artillery spotting location. Russians trapped in the port watched helplessly as Japanese artillery sunk the vessels stuck in the harbor one by one; it was only a matter of time until the Russian defenders were next. The Russians surrendered.

Bringing this military tradition back to planetary invasion, what a planet looks like on its surface is imperative to forging a successful invasion strategy. The force composition needed to successfully negotiate a planetary invasion depends greatly on what that planet has; if a planet has an atmosphere, if it has oceans, vast forests, tundra, flowing volcanoes, jungles, even down to the chemical elements which comprise the composition of its surface. With this variety in mind, it will be unrealistic to build and maintain the equipment necessary for a "general" planetary invasion. While some things will always be required for the spaceborne portion of an invasion operation (training, space transport, supply, and so on), it is clear each invasion will require specific tailoring. This includes specialized equipment and training, environmental protection suited to the planet in question, specially modified air or naval forces for the target, clearing equipment, machines with an appropriately high or low heat tolerance, and so on. While it may be difficult to comprehend for the 21st-century reader of this text, the conclusions are clear: planetary invasion may well require developing a specialized air force, navy, and even occupational national guard equipment *prior* to invasion for use on the target world. These forces can then establish superiority over their pertinent domains, then change roles to occupying security forces once the planet is secure.

Scale is another problem. It has taken mankind several thousand years to cover the Earth's surface, skies, and seas with the military forces we have now. If we were to attempt to conquer an Earth-like planet of approximately the same size, just how much effort—and how long—would it take to build and prepare the specially-tailored equipment and trained soldiers needed? As discussed above, simple mathematics dictate the numbers will have to come from external sources, like human settlements other than Earth, to provide the manufacturing volume and soldiers needed for such an undertaking.

Given the cost of the forces, preparation, training, and specialization which will likely be needed to wrest a planet from an opposing force with the maximum chance of success, the political temptation to succumb to time limits, cut corners, flub production quotas, and brush over the target planet's specific characteristics pertinent to an invasion will be tremendous. Future military leaders should carefully note that past military preparation mistakes could easily crop up in a process as involved as planning for a planetary invasion. Soberly analyzing the operation, complete with a focus on specializing equipment and tactics to the target planet's characteristics is absolutely necessary to maximize the chances of success.

Automated Assault Forces

With the challenges presented to any force attempting an operation on the scale of planetary invasion, it clearly behooves any military force to automate its processes to the maximum extent possible. As Sun Tzu once famously said, the keys to victory lie in knowing the enemy and oneself.[24] When it comes to the human species, we already know that the number of casualties and the relative harm done to air-breathing, flesh-and-blood human forces significantly impacts a human force's political will to continue fighting. Likewise, the ability to continue combat in space will in no small way be tied to the number of casualties absorbed by our far-flung human assault forces.

This begs the question: can humanity really comprehend the scale of casualties needed in order to push on to victory during a planetary siege if it involves humans as the primary assault force? Any operation in history would pale in comparison to the sheer fielded numbers, much less the losses, deployed and incurred during an interplanetary assault.

Obviously, automation provides a partial answer to these concerns. It could one day even completely replace our need to send volunteers far away to satisfy our political needs to use force. As of this writing, automation is undergoing a technical revolution; first in the small-scale military sector via unmanned aerial and land-based robotic forces, then in commerce, and soon enough will be included into large-scale war plans. There is no reason to believe that a mankind capable of interstellar flight does not also possess the technology to craft combat forces made of steel rather than flesh which exist to fight on our behalf. The political concerns associated with placing our sons and daughters in harm's way—which is never easy, in any era—could prevent large-scale human participation in a surface assault. It is difficult to envision a planetary assault completely without human involvement, but it is not difficult to envision said assault with humans taking a back seat and controlling an automated assault force from on high, even from orbit or further away.

While some science fiction glamorizes the combat automaton made in our image, this does not necessarily have to be so; in fact, the shape and figure of any automated assault force should be carefully considered before construction. Much like the assault forces themselves, any automated forces should be specially tailored to the assault in question.

There are two main reasons for this. The first is that because automated combat forces' capabilities are completely under our control during manufacturing, they can be crafted with attributes made to function specifically within a hostile target planet's environment. This means the automated assault force can be built to withstand or operate in differing gravities, atmospheres, and even domains (the enemy's air, seas, land surfaces, or lower orbit). In truth, an automated assault force might be the only feasible method to assault a planet due to extreme environments deemed too hostile for human forces to land, even with biological protection.

The second reason an automated force should be built specifically for a single operation is because these forces can be tailored precisely to the fears of the enemy. With total control of the shape, size, and characteristics of an assault force, the manufacturer can carefully balance assault automatons' aforementioned designs with attributes designed to intimidate and terrorize the enemy. This is an old technique. The Vikings intentionally carved their longships into the shapes of dragons to frighten their opponents; eyes painted on the sides of Greek and Mediterranean vessels were designed to scare enemy fleets. The list goes on, but one important part of terror by design is that it should always be subordinate to the functional aspects of equipment and forces. In other words, a scary weapon which can't effectively fight doesn't stay scary for long. If we are facing a different species, it then follows that they will likely be afraid of different things than we are. Some kind of natural predator found on their planet, perhaps? Does their culture have a taboo or religious prohibition against a particular animal or phenomenon? Can their myths be manipulated to make the assault forces appear as a form of divine retribution or terrifying monster familiar to them? Any irrational weakness presented by the enemy should be leveraged whenever possible for combat success, but will probably require an exceptional understanding of the enemy's culture.

With advances in 3D printing and industrial capacity, the distant future will no doubt raise the prospect of automated and lifeless assault forces and spacecraft which are tailor-made for each mission. This potential niche of the twenty-fourth or twenty-fifth century defense contractor or military-industrial complex will likely find its first efforts here on Earth, during our well-worn and familiar state-to-state conflicts.

When considering including automated forces into a planetary assault force, some things to consider include:

- **Reliability.** Considering the stakes of any invasion and the distances between our bases and our targets, any automated ground force must be resilient in every sense of the word. A force with critical signal weaknesses or power concerns which could cause them to suddenly switch off or fly out of control during battle should not be fielded.
- **Terror.** Shark teeth are not painted on the noses of A-10s for nothing. Like it or not, intimidating the enemy is an excellent tactic and should be pursued in the most practical way possible. After all, there is more relative freedom for design choice and appearance when crafting a war machine from the ground up. Does our foe fear a particular animal native to its world? Perhaps automated forces can be fashioned in the image of one of our foe's native natural predators? Have interrogations with prisoners provided information which gives us clues to its fears? A frightened foe, while unpredictable, is more easily beaten; and friendly forces seeing a fleeing enemy are that much more emboldened.
- **Function.** Perhaps most significantly, understanding the limits of current automation is critical to employing it correctly during the conflict in question. There are just some things an automated force will never be able to do by itself—develop reliable strategy consistently, understand and manage the needs of biological forces, complete certain maintenance functions, interpret vague orders, and seize the initiative when it presents itself. These things are best left to human forces to do, but in conjunction with and supplemented by automated forces.
- **Control.** To what degree of freedom should automated soldiers be granted? This depends largely on the technological progress of AI, but it is still important for human commanders to maintain some kind of control and

authority over any orders the machines issue to themselves. How far and how fast they can go could prove both a blessing or a curse in any tactical situation, and a human commander should be able to overrule an automated force which may be careening out of control.

- **Environmental concerns.** If automated soldiers are to be sent in harm's way, one ought to give them the best preparation possible. This means carefully crafting them with as much adaptable characteristics as possible for the targeted planet in question, even to the point that the forces could not survive or function on Earth. Is the atmosphere caustic? Then additional work on corrosion-resistant alloys is worthwhile. Crushing gravity? Nothing a reinforced titanium skeleton won't fix. An oceanic planet? High winds? Extreme temperatures? Operational considerations must be incorporated into automated soldier design, even if it means never being able to use that particular soldier for another assault again. Since the costs associated with any planetary invasion campaign will be excruciating for any state, and since travel times in space will likely prohibit encore attempts, assaults will only get one good chance to succeed. This means the desire to maximize the chances for success *the first time* will necessitate using tailor-made war machines to the maximum extent possible.

In short, planetary invasion and combat in the future will bear little difference from terrestrial military campaigns, in the logistical sense: travel, distance, attrition, troop choice, quality of personnel, reliability of equipment, veterancy, and luck will continue to affect military operations on other worlds and planetoids just as they always have on this planet. Technology, while always a critical component of any combat operation, will continue to confound as well as inspire assault force planning. Any political unit attempting a planetary assault would be wise to first examine themselves, the enemy, and to carefully consider the gargantuan logistical challenges associated with interplanetary warfare before committing to such a high-risk venture.

3

Ideological Factors

> It is when one side goes against the enemy with the gods' gift of a stronger morale that their adversaries, as a rule, cannot withstand them.
>
> —Xenophon

Success in war and conflict is often a question of attitude. How a species or culture prepares itself for elongated conflict with ideas and discussion is just as important as how it physically prepares with fleets and weaponry. Mental preparation directly contributes to the length of time at which a primary political unit can reasonably expect to continue fighting. As the Carthaginians discovered in the 3rd Century BC, a Rome which remains on its feet even after multiple defeats, and possesses the will to continue the war, is not a defeated foe, despite what warfare's customs of the day demanded.[1] Rome simply prepared itself for victory at any cost—and that is not an easy thing to do. No doubt there were those in the city willing to give into the Carthaginians, but since defeat meant death and enslavement, these voices were never heard.

If we thought terrestrial wars were long, lasting years and costing thousands or millions of lives, wait until we experience an interstellar one. Deep space warfare will both present a high intensity conflict when battle is joined, and a comparatively slow waiting game as forces maneuver and travel great distances to conduct their bloody business. The time required to cross vast distances between planets and stars could easily push the duration of wars into generational conflicts, which will require a completely different mental attitude than we have now. It is worth examining the non-physical ramifications of interstellar and interplanetary warfare, and the best way to prepare for future conflict in space.

The Will to Fight

Morale is a tricky thing. Also termed the enemy's "spirit" or *élan* depending on the source, millennia of warfare here on Earth has taught us that how combatants feel plays a tremendous role in combat. Prussian military theorist and combat veteran Carl von Clausewitz notes "military activity is never directed against material force alone; it is always aimed simultaneously at the moral forces which give it life, and the two cannot be separated."[2] Hostile feelings, the desire to be seen as brave, envy, pride, passion, experience, and a variety of other factors combine to form this concept of morale.

There have been many battles where a beleaguered foe has quit the field after a variety of tactical-level events ended up going the wrong way—their general had been killed, their forces had been buffeted by storms or horrific weather the previous night, supplies had run too low to continue, relief had never come. After this bad luck, the enemy army had but to take the field and deliver the final blow, sending the enemy scared and scattered to the rear. But on a larger scale, these tactical vignettes which claim how morale can be cracked to achieve victory are deceptive. Our experience in warfare indicates morale is a foolish target in and of itself. Like a superstitious gambler who thinks the next hand will set him right if he prays to the right god, many generations of military strategists have pursued an enemy's morale as the ultimate in seductive strategic red herrings. Give it the right whack, and the enemy's entire resistance, an entire nation's will to fight, could crumble into terror-induced sobbing and despair. This hope has been disproven in every single modern war; the will to fight is simply too durable to destroy through physical attacks alone.

Moreover, humanity's ability to predict the effects of striking an opponent's will to fight has been woefully poor. Many a commander has slaughtered large numbers of non-combatants, squeezed cities into submission through starvation, or desecrated a culture's places or items of precious cultural or religious significance in pursuit of the magic formula which would break the enemy's will. In most cases, these commanders only end up with an enraged and hardened foe willing to fight even harder. With the advent of modern airpower, humanity finally and painfully realized searching for and physically assaulting the enemy's strategic morale is nothing more than a wild goose chase. Early airpower theorists like Giulio Douhet thought an unrelenting pounding from the air would surely force an enemy to scurry for their white flags and cause their cities to capitulate as fast as their mayors could send the surrender message.[3] He was wrong; as we learned in World War II, blasting an enemy's cities and civilian population only stiffens resolve.

British Air Marshall Sir Hugh Trenchard, along with his American counterparts, believed the enemy's morale was the principal target in war and could be broken through relentless offensive air attacks.[4] As 1918 dawned, Trenchard could only look upon the scores of downed Royal Air Force fighters and lost airmen, whose training and production cost exceeded the damage done to its German foe, and note how German morale was curiously untouched.[5] While air power can make an enemy miserable, by itself it cannot demolish morale.

What we have learned, then, is that destroying an enemy's morale and will to fight *cannot be* an objective but instead is in fact a *bonus effect*, caused primarily by one major factor: the specter of defeat. Future interstellar campaign commanders must understand this. Put another way, winning by destroying the enemy's centers of gravity and pursuing *operational* victory is the best way to systemically eliminate an enemy's morale, while pursuing *strategic* victory should not concern itself with enemy morale. As defeat looms closer and begins to dawn on an enemy population and leadership, as their troops stop coming home, as their daily life becomes impacted in worse and worse ways, only two possible reactions can happen: either the culture doubles-down on their wartime efforts in the face of certain annihilation, like Japan's civilian population during the final days of World War II, or pressure to end the war before it gets worse increases in lock step with the defeats suffered. The latter is only possible when the war demands of the victor are less than unconditional surrender. Seeking total victories is the best way to remove morale as a factor which could hasten victory; after all, what is the incentive to surrender when a civilization and culture is facing the possibility of complete destruction? Only total national exhaustion in the face of impossible odds will compel an enemy facing unconditional surrender to lay down his arms.

Besides entire societies, armed forces themselves also possess morale. Morale is difficult to read beyond generalities, but a good commander always tries to keep his fingers on its pulse. Clausewitz notes one "should take care never to confuse the real spirit of an army with its mood," and this is good advice which applies to spacegoing forces.[6] Mood, after all, is highly relative; an individual soldier or sailor can have a lousy day, completely independent of how his unit or state has performed in that day's struggle. No doubt there will be some long, boring days aboard spacegoing vessels as the forces onboard carry out their long voyages to destinations which may be equally boring. While this affects mood, this may not necessarily affect morale. Morale is better explained as the overall impression the average trooper has of his or her unit and its ability to perform. Clausewitz explains morale is a critical part of the "military virtues" of a unit, which includes discipline, drill, skill, and pride.[7] Taken together, morale then appears to be a rather shaky thing; a heavy blow on one or more of these things can cause the whole virtuous combination to come crashing down, "like a glass too quickly cooled."[8] Clausewitz cautions these factors can be overrated, and when challenged or upset lead to a "boastful pride" which hides a lack of confidence the armed force has in itself.[9] Spacegoing forces, which will require a high degree of discipline to be kept in check over long distances and periods of inactivity, should take note "grim severity and iron discipline" do not hold a force together by themselves.[10] Morale, in fact, may be more frangible than the vessels which carry their hosts through the stars. Only the future's "tempered, battle-scarred" space warriors will be able to tell us the truth.[11]

Facing a Non-Human Opponent and Its Repercussions

We are fairly certain that the first foe we will face in the cosmos will be us. In this regard, the human race has ample experience blasting each other into smithereens, and this experience will no doubt be happily employed by future combatants. However, it is worth discussing what would happen if, perhaps when, we face a non-human opponent, and how this relates to how we think about war.

Our Species' Reaction to a Non-Human Competitor

Initial reactions to encountering non-human sentient beings will likely be similar to encountering animals on Earth for the first time; a blend of curiosity and cautiousness, perhaps with a pinch of superiority. If what happened between our species and the species of other earth-evolved hominids as found in the fossil record is any indication, we can expect to treat any non-human adversaries we happen to meet with aggressiveness from the outset.

Luckily, it is not necessary to wait the required thousands of years to evolve into a species better suited to a multi-race galaxy. It is necessary, however, to prepare ourselves for meeting our first extraterrestrial sentient competitor and the concomitant potential for acrimonious conflict.

There is good reason to believe we will have to repress a sense of loathing, of unnaturalness, or even of intense fear or hate upon encountering something so different than what we know life to be, and then seeing it do things like us. Indeed, not every sentient species is guaranteed to be humanoid in appearance, unlike those on episodes of *Star Trek*. Rather, statistically speaking there is probably a greater chance the sentient species we first meet will be something other than a mammal, and perhaps not even an animal. After all, large and intelligent reptiles existed on Earth much longer before we mammals came to the fore; what could they have achieved if the Earth's climate had not changed naturally, nor had been struck by an asteroid now at the bottom of the Chicxulub crater?[12] While plant life as we know it implies a sedentary and immobile existence, must it be so everywhere else? What is the definition of inorganic life, and is it theoretically possible?

The method of first contact with a sentient competitor, if controllable, will have to be carefully managed. One assumption, though difficult to understand, should be made about an extraterrestrial competition: everything about an alien spacefaring species' thinking, expectations, actions, and perceptions must be assumed to be completely different than ours. This makes contact hazardous, confusing, and slow. Even the best pace at which our species' first ambassadors (willing or otherwise) should begin their introductions is a guess, since even our scales of passing time are likely to be completely different.[13] No method of communication can be guaranteed to work. Many science fiction works focus on communicating initially in the language of mathematics and science, which is as good a start as any since presumably a strong science background would be required to have reached the point where interstellar travel and first contact could be made in the first place.

Nevertheless, prime number repetition and basic chemistry or physics must eventually give way to negotiation and complex communication. The time required to do this may outstrip any patience either species may have. Humanity can best prepare itself for first contact by understanding not all of us are as interested in space exploration nor as friendly to other competitors as we might think, and are perfectly happy to live in a universe without ever meeting another species.

The key to any introduction of a single member of humanity to another sentient spacefarer, even long after first contact, is contact management. Spending time studying appropriate emotional management techniques designed to prepare someone's psyche prior to encountering another sentient species in person would be time well spent. Essentially, preparing to meet another sentient species will be like preparing for one's first visit to a particular foreign country, at least in basic principles. Unfortunately, something we all depend on during such foreign forays will be conspicuously absent: the human factor. This refers to all human travelers sharing the universal human condition: gestures, body language, family ties, food and drink, clothing comparisons, laughter and humor, and so on. In other words, humans are not prepared to encounter species with which it shares sentience but has nothing in common culturally, and cannot therefore rely on baseline human behavior (which will make first meetings terser than we expect). Not even that last resort of the desperate traveler, body language, will be available.

The Power and Primacy of Fear

In his legendary work *The Peloponnesian War,* author and historian Thucydides observes people go to war for three main reasons: honor, fear, and self-interest.[14] Of these three, fear permeates all three situations. While it may not be the causal factor of every conflict, in any decision to go to war and risk the nation's existence, a degree of trepidation is certainly involved. As we shall see, space's unique environment and the challenges we will find there make a strong case that fear is here to stay, and exerts an oversize influence on interstellar conflict.

What can a two-thousand-year-old conflict teach us about going to war in space? Mainly, it reveals to us the classic political conundrum—which will certainly extend beyond humanity—called the security dilemma. The Peloponnesian War began as two political entities and their allies, Sparta and Athens, began to fear the other's military power and influence. Sparta, the acknowledged superior military power, eyed Athens' economic gains and growing military and political power with suspicion. Athens, the rising power, viewed Sparta as attempting to keep Athens' growth in check, and in turn preserve their own preeminence. As doubt and mistrust crept into the relationship, diplomacy failed and Athens took steps to violate a treaty which kept a shaky peace in place for the preceding twelve years.[15] War was on, and war is always risky.

This situation is sometimes referred to as the "Thucydides Trap" after the same Greek author. The "trap" refers to the fact that in pursuing their interests, both Sparta and Athens could not avoid war; in doing what was best for them, a behavior which normally grants success and security, they were forced into a situation where fighting was the only solution. The Peloponnesian War is often evaluated by military historians and scholars who seek different lenses through which Athens and Sparta may have viewed the brewing conflict. Besides security, differing economic systems—a cosmopolitan Athens dependent upon trade, and an agrarian Sparta which depended on slaves—is often cited as a reason both cultures collided in such spectacular fashion. But at the end of the day, power is what forced them to fight, and fear is what also made power the object of the fighting.[16]

The security dilemma is often cited by international affairs scholars as the default situation between primary political units. It describes how one political entity is never entirely sure how secure it is when facing another independent political entity because one can never be too sure about the true power potential of the other—one can never know just how much security they need to deter a neighbor of unknown strength. A weapon which one state claims is only for defense can be used to attack. This insecurity causes a natural inclination for a state to arm itself to a level which "feels" superior enough to either deter or crush any attempt by a rival state to overcome it by force. This act of arming itself, however, unnerves other neighboring states, whose limited ability to analyze its rivals forces it to arm itself. Even if a rival is truly arming only for defensive purposes, and even if they make oaths to the same and all national behaviors indicate defense is all they are after, there is always a kernel of doubt; a pinch of mistrust; a "worst case scenario" analysis where a well-armed state suddenly, perhaps overnight or with a leadership change, decides to use its acquired arms and military power to attack its rivals rather than console itself solely with defense. This in turn forces rival states to arm themselves to a level they feel will balance the threat they face; which feeds insecurity in the first state, forcing states to stretch their arms acquisitions to their economic limits. All the while, fear is a constant companion.

This is an old tale, and very human. But this brings us to a critical question: will it be like this in interstellar conflict? If the combatants are only human, the answer is a resounding yes. If anything, confusion will be enhanced by the vast distances between rivals, not to mention slow and incomplete intelligence which comes with this situation. Land and territory, expressed in space as the stellar system, are still important to the human race, and we will no doubt fight ferociously to defend what we think or perceive to be ours. In other words, humankind will no doubt face the same security dilemmas in space which we deal with on Earth; the only change will be the setting.

We may be tempted to think other sentient spacefaring species, should we ever encounter them, may not have the same concepts of territory. They may not have a history of endemic warfare focused on territory such as we humans do, which is, after all, a primate characteristic. Could we potentially someday encounter the stellar equivalent of the mythical stereotype of the Native North American—no concept nor need of property rights, interested only in endemic warfare as a means to prove themselves as brave, rather than a conquering victor? Could we meet a fellow star faring race with no need for modern economies, industrial ambition, scientific discovery, nor a need to pillage the interstellar landscape?

Not likely. If a species has survived long enough to develop sentience and consciously decide to escape its homeworld prison in order to venture into the galaxy, there is little chance it would never have experienced fear. And if a species can experience fear, then the security dilemma will be a central, if not the primary, tenet of its foreign policy. Why is this so? Fear is not just a byproduct of evolution; fear is also a survival mechanism, and a very important one at that. Fear causes an individual to imagine the worst-case scenario when he sees the bushes rustle; when she sees a volcano smoking in the distance; when he encounters a hole in the ground; when she sees her loved ones wasting away from disease on their cot. Fear makes organisms conservative; it forces them to think; it makes them vulnerable to confirmation bias (discussed in Chapter 7) in a way that temporarily cripples the species rationally, but grants them the behaviors to survive in a hostile landscape.

In their early years, fear keeps a sentient culture and its political units alive. Indeed, many customs throughout the world which survive today and which are firmly embedded in cultures are nothing more than good practices borne of fear or tragic experience. Culture, in many ways, is simply a local recipe to arbitrarily execute procedures which every human needs to survive. Cultural habits between civilizations which address day-to-day existence often exhibit great variety on a very small list of topics and behaviors, which gives credence to the notion that all members of our race faced the same problems at one time or another, and faced them while afraid. Fear is an excellent source of custom.

Fear is also a tool—it allows people to execute heroic feats to protect their families or defeat enemies, and at the same time provides the necessary fuel to cajole statesmen into going to war. Fear, and its effect on our species, is a sad and ultimately inescapable part of being a primate, and will accompany us to the stars. It is highly likely we will share this burden with other spacefaring sentient species.

The Hazards of Disunity

Political systems are under no obligation to enter a conflict united. In fact, if history is any indication, quite the opposite is true for most of humanity's existence. Deep space warfare carries with it implications of planetary-wide or species-wide political agreement to go to war, a will to continue fighting, and a unified effort to supply, proctor, and conclude a conflict.

It will matter what political condition humanity finds itself in during its first interstellar conflict. How we as a species are oriented towards potential sentient extraterrestrial life if and when we encounter it will make a large impact on how wars could develop. Below are a few potential conditions to consider which could affect the overall human war effort, should it come to that.

Xenophilism versus Xenophobism

Extremes are never reliable. Sadly, an extreme reaction to the existence of a sentient, foreign, and competitive space faring race is likely upon first contact. This is more due to what we know ourselves to be like than to a logical or rational computation which attempts to analyze and predict what our reaction *should* be. In truth, we are primates—and most primates attach particular and special value to control and monopoly of the use of force, wealth and resource distribution, sexual rights (perceived, legal, or otherwise), fairness and justice, and security. While there is no guarantee the first non-human sentient species we could encounter will also value these factors, we already know we do; and humanity will, consciously or otherwise, instantly size up our new competition within this cultural lens. It will be an inescapable fact that hours after first contact is made, national security councils and advisors to leaders around the world will begin huddling in their most secret of rooms with their most trustworthy supporters for the express purpose of analyzing the new threat in terms of security. To do otherwise would, from a military perspective, be foolish; and it would also fail to understand who we are as a species.

Ultimately, what we do as a species will depend greatly upon the reaction of the public at large to the discovery of a new and similar spacefaring sentient race, the political unity of the planet, and the political willpower of the great powers who stand to be able to do something about the discovery of our newest neighbors. Our chosen strategic courses of action will rest somewhere within a spectrum which stretches from xenophilism to xenophobism.

Xenophilism

Put simply, xenophiles wish to get closer and interact more with people or cultures—or sentient species—not their own. This attitude has basic but clear roots: the natural curiosity, inquisitiveness, and desire for understanding which all humans innately have. Anthropologically speaking, this desire to learn and understand new things is an evolved human trait. The more an organism understands about its environment, the better it can pass on this knowledge to the next generation, the more successful at survival it will be, and the greater chances of success in intra-species competition it will have. The same evolutionary prerogatives that force us to closely watch an adversary also encourage us to trade with them, send diplomats to their court, and establish cultural exchanges.

The word "xenophilism" (and its opposite described below) is a Greek word meaning "love of outsiders." Of the two extreme ends of the spectrum of choices we will have for how we address future extraterrestrial contacts, true xenophilism is probably the least likely to actually happen. While there will certainly be plenty of people who will be interested in becoming as close as possible to any newfound galactic neighbor, these people will not likely be in government, nor will they likely be decision makers.

A measure of prudence clearly reveals the problems associated with unrestricted xenophilism, and why totally unqualified fraternity with an alien outsider is not realistic. Our natural inclination to meeting anyone whom we do not know is suspicion and trepidation; as it should be. This comes not only from genetic predispositions which have helped our ancestors survive, but also from millennia of experience by statesmen who know it is much safer to cautiously build a relationship with foreign entities who have unknown priorities rather than running open-armed into their embrace. Indeed, it is best to see any future encounter with extraterrestrials as a meeting with competitors and not purely as friendly fellow travelers who simply happened to encounter each other along the roadway.

In the end, a nominal xenophilic attitude, combined with a reciprocal one from the encountered extraterrestrials, can be useful. The most useful part of this attitude is its ability to avoid immediate conflict and violence upon first contact, which is a naturally volatile situation. Other useful factors include meeting an extraterrestrial with a mind for discovery rather than for conflict. Discovery brings its own rewards, which includes the all-too-human thrill of learning about a new species and culture for the first time, technological and information exchange, and cooperation in interstellar exploration. Because of space's vast expanse, there is no reason to believe conflict is inevitable simply upon discovering the existence of a fellow traveler.

Xenophobism

While xenophilism is useful for many positive reasons, conversely xenophobism is useful primarily for security's sake. This explains why the vast majority of current statesmen, and every single one of the most historically successful statesmen, adopt an attitude of mistrust and doubt in the foundation of their foreign policies. Recall what we learned earlier about the anarchic environment of international relations: since there is no overriding authority above the primary political units in a system, this doubt and mistrust is natural and expected.

If xenophilia can be described as optimism, xenophobia is its dour antonym, silently insisting to the statesman that "pessimism may not be pleasant, but it keeps us safe." Put briefly, where "xenophilism" is the love of outsiders, "xenophobism" is the fear of others. Note xenophobism is *not* "hatred of outsiders," as our immensely logical Greek forbearers understood it is irrational to hate something one does not yet understand. To fear, rather than hate, the unknown is very much the human condition. This definition is easy to understand; xenophobic cultures tend to shy away from foreign intercourse, participate less in trade, and limit foreign contact to the bare minimum their political system can accept.

Without a doubt, a xenophobic attitude's worst consequence is how it nudges opinions towards suspicion and violence. A xenophobe tends to mistrust and doubt foreigners; everything they do somehow relates only to treachery. Extreme xenophobes are difficult to reason with, and conducting state-to-state business is nigh on impossible. In a xenophobe's eyes, every official visit is seen as a calculated action designed to case one's home and country for future foreign exploitation; each word from a foreign dignitary is a honey-dipped lie. Perhaps not unsurprisingly, xenophobism is perfectly functional in international relations. Indeed, sometimes official visits by foreign delegations really *are* grandiose schemes to spy on a rival, all while under diplomatic cover. While it may be considered "nasty," xenophobism is neither always out of line nor completely incorrect, especially given historical examples of diplomatic trickery and deliberate state-to-state misleading. Even though humanity has not yet met a non-human power, we already know this play will certainly be reenacted upon a stellar stage; power politics, regardless of setting, provide no other alternative. Therefore, the best aspect of xenophobism, and why it tends to be adopted by many a successful empire here on Earth, is that it keeps a state safe until a threat can be effectively analyzed.

Where discovery is the watchword of the xenophile, security is the corresponding xenophobe's marching orders. Xenophobism does not necessarily remove cooperative and productive interchange with another political unit, but it certainly delays it and often restricts it. Again, this is not always a bad thing. Where xenophobism gets cultures in trouble is its tendency to encourage civilizations to remain distrustful too long, and squanders valuable interchange opportunities or poisons a relationship beyond the point of recovery.

One of the most striking historical xenophobia examples is pre-modern Japan. From approximately 1600 AD until 1853, Japan had succeeded in both unifying its disparate political entities into a unified geopolitical nation, and in keeping non–Japanese powers largely out of Japan's geographical territory. In a unique national effort, Japan's leading political entity, the Tokugawa Shogunate, executed a misinformation and physical restriction policy called *sakoku*, or "closed country."[17] Few nations since have successfully undertaken such an effort. The populace was instructed that foreigners found illegally in Japan (which were just about all of them, save a small Dutch enclave off the coast of Nagasaki harbor), even shipwrecked or lost sailors, were to be denied services, port access, and depending on the situation, killed on sight. Nefarious Western influences, especially Christianity, were blamed for making trouble and violently stamped out of Japan. Those caught with forbidden knowledge or worshipping foreign religions were punished with ruthless efficiency by the Shogunate. For nearly 200 years, with the exception of the aforementioned Dutch post near Nagasaki, the Shogunate successfully executed this policy.

We can learn from this policy by what happened to Japan. In the end, Japan's quest for perfect security and cultural hegemony was nothing more than a political screen to keep the Shogunate in power by controlling Japan's foreign policy. The Shogunate's domestic political rivals, other clan leaders, were all too happy to trade with foreigners if they could and try to gain a material or wealth advantage over the central government. In fact, the *sakoku* system was merely what noted Japanese historian Kenneth B. Pyle describes as a way to "enhance the legitimacy of the shogunate by resolutely bringing foreign relations under its control."[18] In Japan's case, if permitted to grow in strength and influence the Shogunate's rivals could appeal to a foreign power to obtain political and military assistance against the Shogunate. The Shogunate's fears were well founded; this is precisely what happened after Commodore Matthew Perry's expedition to Tokyo forced the country open to trade and exchange with the West.

The lesson is this: Japan held onto a xenophobic policy too long, despite every indication and unfounded belief that it was working to help keep the Shogunate in power. When the Shogunate brought its head out of the sand, it encountered a world where its swords were obsolete and powerful cannon-armed vessels belching smoke were the new normal. Since Japan was never completely cut off from the outside world, the Shogunate knew of the existence of these things; they just never imagined they could menace Japan. All these vessels now seemed to be bearing down onto Japan at once, with seemingly every nation smelling the blood of a hopelessly outdated feudal throwback. Japan avoided becoming a European colony only by the skin of its teeth, and had to fight four wars to do so.

Thus, we can see the lesson: while xenophobism is natural and at times even prudent, like every other foreign policy it lies on a spectrum, and has its limits.

Our orientation towards foreign interstellar relations will likely include degrees of xenophilism and xenophobism rather than exude an attitude purely one way or the other. This will depend greatly on which government is in charge at the time, which government (if the planet is not politically unified) actually makes first contact, and the prevailing attitudes of public opinion at the time of the encounter. Speed and secrecy will play large roles in how the public perceives a first contact situation. Even to a casual observer who is uninterested in the issue, news of any first contact generates several questions all at once: who are the visitors? Do they look like us? Are they friendly, or otherwise? Are they interested in trade and interspecies discourse? The answers to these questions, while important, are best answered first by competent authorities with military advice. Only after security is reasonably assured can it be safe to unleash the uncontrollable knowledge of the existence of a non-human sentience upon an unsuspecting and unprepared public—along with a surely frenetic and noisemaking media.

Political Divergence in Populations Outside the Homeworld

In a situation where humanity has reached for the stars but has yet to encounter a race different than its own, unity will still be a problem. We will still face the same troubles when our own fellow humans are our primary competitors just as if we were competing against an alien race. To be sure, any planetary settlement, any outpost with a population, any inhabited area beyond Earth could presumably politically diverge from its founder's views at any point. Any place inhabited by humans in space will no doubt be perilously distant from its legitimate and effective authority, and once initial concerns about survival and security matters are over, political trouble is sure to appear. A distant and relatively less-governed holding will likely have plenty of time for human organizations to grumble about their interests being neglected by a central authority, and still more time to allow grass-roots political opinions and movements to take hold.

We know this could happen because it already has. Consider the example of the British colonies in America. The colonies, following the first successful permanent settlement in Plymouth in 1620, grew steadily in strength both via immigration and royal patronage until 1763. The British government, busy with many commitments in Europe and across the globe, hit upon a clever policy to maximize colonial growth and minimize British state support. British Prime Minister Robert Walpole consciously instituted this policy of "salutary neglect" in the 1720s.[19] For nearly 150 years, the colonies were more or less allowed to develop their own culture and local organization, provided they were deferent to the British state and continued to participate in Britain's mercantilist system by depending on Britain for finished goods as the colonies' primary trade partner.[20] This began to end in 1763. British authorities began to levy greater and greater taxes upon the colonies to pay off its war debt from the Seven Years' War, ending salutary neglect.[21] The state, failing to recognize (or ignoring) the colonies' political, economic, and social divergence, reminded them of their duties to the crown through force, and inadvertently catalyzed the United States' war for independence.

Planetary settlements will face similar relationships with their home-world masters, though the end does not have to be as it was with America. Future states must recognize how settlements become independent and then place themselves in a position to stop or mitigate unwanted independence. With a healthy dose of state security forces, financial support, and with a minimum of arrogance, states can keep extrasolar settlements and governments united. Planetary settlements must be treated as the wonders they will become—a group of intrepid humans, defying the terrifying hostility of space, who plunge themselves through the starry blackness to a world previously devoid of sentient life; master the flora and fauna; become familiar with the new land; interrupt local evolution to bend it to their will; survive local weather phenomena; and ultimately adapt to a foreign world they can come to call their mother. The first time we do this it will be truly remarkable—but we must first begin by understanding what could happen politically.

Planetary Unification: An Impossible Dream?

A futuristic scenario involving spacefaring forces begs the question: for whom do these forces fight? A nation? A cause? A united world? The answer to the question involves a complex algorithm of several competing factors, but ultimately comes down to one word: politics.

The political objectives of the primary political unit—be it a nation, a settlement, a unified planet—before executing interstellar combat will dictate who goes to war and for whom. It is entirely conceivable that humanity could take to the stars well before we have worked out our problems on Earth well enough to be considered a "unified" species. "Unified," after all, is a highly relative term. If this is the case, we should expect our disunity to lead to the same political structure we have here on Earth: a Balkanization of territorial claims, empty space, and political competition and misunderstandings. While planetary political unity is theoretically possible, there are many reasons why it will be exceedingly difficult.

The Hazards of Rapid Unity

Consider a situation where the entire planet is politically unified into one central and functioning political bloc. As science fiction often preaches, this is a good thing. Right? Well, it depends. Politically, a unified planetary government is unrealistic for several reasons, and in any case a rapid unification could be just as dangerous as arriving at where we seem to be heading: a loose and self-interested collaboration of space faring nation-states.

First, it is not clear why a unified planetary government would make the deliberate decision to become a spacefaring civilization in the first place. It seems clear that a certain amount of political maturity, technological and financial support, and political stability is necessary for a species, as a unified force, to take to the stars as a deliberate course of action.

Second, a rapidly unified species is by definition politically fragile. The word "rapid" implies unification occurs before all parties are ready, and this obviously leads to unpredictable domestic consequences. Even our most long-lasting current diplomatic and political arrangements are tenuous at best, and held together at the mercy of their independent political parts. Political unification on any level resembles a business arrangement, and not family ties; history has proven their nature is ephemeral. Even history's longest lasting and deeper partnerships—the Anglo-American Alliance, for example—require a great deal of political tradition, cultural commonality, and shared security concerns, along with a healthy dose of trust, to keep it alive. Given the disparity of cultures, disagreements, and conflict here on Earth, planetary unity is therefore unlikely: groups of humans do not happily sever their old sovereignty and join a new and untested political band without developing serious security concerns. In short, political unity requires both a reason for its existence, usually for security, and then constant maintenance to sustain it. If at any time any of the members party to such a unified entity wish to leave, the unified political entity by definition no longer exists.

Sovereignty, then, is the obvious crux of the matter. In Western statecraft, there is little that exists which is more precious than sovereignty. States, political units, and the world has yet to see a serious political challenge to the state system set up by the Treaty of Westphalia in 1648. But sovereignty can also become a weakness. A stubborn insistence to adhere to sovereignty concepts will keep humanity Earth-focused, since states and their territorial claims are the heart of sovereignty.[22] Without appropriate updates to face the realities of our civilization in space, planet-wide political unity of Earth will remain a dream.

This is because future political units on a regional, continental, or planetary scale will invariably require states to surrender their sovereignty for the good of a group. There is no other logical conclusion to make when considering what is necessary to politically unify an entire planet. Indeed, surrendering the sovereignty of the disparate nations which dot Earth's surface to a planetary-wide political entity is the *only* precondition to planetary unity.

For all intents and purposes, sovereignty is technology. It was invented by humans—some say imagined—and therefore could potentially be unmade, or at the very least reformed. To be more precise, the concept of sovereignty is simply an extension of primate priorities and addresses things primates value. Sovereignty is an expression of territory and authority, deemed necessary by our species for security and backed by the use of force. If a suitable replacement political system could be developed which satisfies our needs for territoriality, security, and control, and simultaneously allows for planetary unity, there would essentially be no obstacle to planetary political unification.

As daunting as it sounds, though, political unity is not an impossibility—provided it is done in a well-paced and deliberate manner, there is no reason to believe the tremendous forces of security, technology, and economic development could not someday provide the right conditions for political unity on a planetary level. Before that day, though, states will somehow need to be convinced that giving up or unifying their sovereignty, and therefore outsourcing their security, is in their best interest.

Ideological Unification Is Never Complete

Even if we one day achieve political unification on a planetary scale, humans have proven themselves incapable of agreeing *en masse*—just ask a group of random strangers to agree on pizza toppings and you'll see a fascinating political game play out, complete with bickering, competition, and accommodation. It is entirely feasible for a nation or group of people to be politically or economically motivated enough to head into deep space to pursue their military and economic goals without ever politically agreeing with everyone they meet.

As discussed above, governments are going to be the entities making the big space exploration, security, and military decisions; and these governments have ideologies central to their existence. It would be foolish to think these ideologies will disappear upon completing a successful space-based project; rather, these ideologies will likely be strengthened and displayed by governments as proof of their effectiveness if their difficult, expensive, and risky space-based projects succeed. If a government were to successfully establish a planetary settlement, for example, there is little doubt that government would impose on that colony mandatory membership in its ideology as a condition of its existence and protection. In other words, a democratic nation founding a settlement on another world would expect those settlers to also be democratic. To do otherwise would be counterproductive and wasteful from a political perspective.

Success in accomplishing space projects under the auspice of one state or another will in fact end up delaying planetary political unity rather than accelerating it. Politically, it would be silly to think the mother government in our above example would somehow abandon a project like a planetary settlement after expending so much energy and resources on establishing it. The average citizen here on Earth, baffled and awed by the effort and skill displayed by the intrepid settlers and by the government's efforts to succeed, would likely see any successful establishment of national power in space as proof that their nation is truly great and unique, thus strengthening home nation ideology. This, in turn, would further delay some kind of agreeable "planetary ideology" or unified government if one was not yet established. In other words, if we begin seriously exploiting space under a sovereign state structure, our creations in space will likely reflect that situation for a very long time.

Leaving aside space for the moment, here on Earth we know we regularly disagree with one another. There are plenty of folks who would disagree with space projects in general on principle; apportioning vast sums of our wealth to what they would see as wasteful or overly grandiose space adventures is a ridiculous idea to those who prefer to focus on the serious problems remaining for us to solve here on Earth. Religious disagreement, and perhaps cults, will likely be a continuous problem as the fog of the known universe continues to recede, providing science with more answers to questions previously handled exclusively by religion. Further, humanity has a unique problem: even as a tiny minority of our wealthiest citizens chomp at the bit to ride an experimental rocket into lower-Earth orbit, the majority of our race remains fractured by ancient grievances, poverty, and tribalism. The irony should not escape the reader that the very year space tourism will begin, several billion people on this planet will still have poor access to clean water and medicine, lack security, defecate outside, will never drive a car or ride in an airplane in their lifetimes, and will likely never even see a computer. The difference between our greatest accomplishments and meanest existence is striking. When Earth states finally do reach the stars and attempt to outdo each other in space as they surely will, what percentage of our luckless comrades left behind by their lot in life will even understand what is occurring, let alone be in a position to politically support or obstruct it? It is very possible entire states in the developing world will be left completely out of space competition due to the unbelievable expense required to participate in the great games beyond the ionosphere. Much like the modern system of warfare, they will simply be unable to compete. A disenfranchised race does not bode well for planetary unification.

Three Potential Pathways to Unification

Given all this, planetary political unification would still bring benefits to humanity. By definition, political violence would likely decrease; and while there are always those willing and able to fight with current authorities, a government on a planetary scale would have the resources available to solve a myriad of organizational and existential problems. While it would not be perfect by any means, a planetary government would probably be a step closer to perfection than what we have now. Thus, we owe it to ourselves to discuss what political planetary unification could actually look like.

As it stands now, planetary unification could probably only happen via two basic human requirements acting as its impetus: adequate competition to drive the need to unify, and sufficient technological development to support the vast communication and military response network to support and defend such a planetary-wide organization. Only the insatiable human need for security is strong enough to act as the required incentive to politically unify our species on a planetwide scale, and this security could never be guaranteed without the capability to quickly and reliably communicate with each other. Responding to any perceived threat in a timely manner to make a military difference is also a major problem.

The first requirement, technological development, will likely happen independently before planetwide political unity, rather than the latter pushing the former. With the 20th century's last two great gifts to humanity, the internet and global telecommunications, one can begin to see what reliable planetwide communication networks look like. Still, there is a long way to go in order to ensure reliability, connectivity, and communications security. Successive generations will look upon our once-mighty 56k dialup modems and current 4G internet architectures as archaic; as of this writing, the impending

5G "revolution" stands to blow away previous speed and connection reliability concerns. Technological progress like this is a necessary prerequisite to sustain any kind of planet-wide deployment and military logistics capability.

The second requirement, that of a rapid military response anywhere on the globe, is a tricky thing to achieve. A planet-wide mobile military force's first objective will be to deter states; there is no sense picking any fights if a powerful military force can suddenly show up on your doorstep. Providing physical security to all participants of a global political hegemony, while secondary to deterrence, will still face an uphill battle. Guarding seven to twelve billion people is no mean feat. While the U.S. Air Force has pioneered a concept called "Global Reach," and is capable of rapidly deploying military forces around the globe within hours, the scale of its deployments is still very small compared to the sheer size a global stability force would require. Today, only the U.S. Air Force is capable of rapidly deploying any force of any size in this manner, and could serve as a template for a future rapid response planetary security force. This, of course, assumes the hairy problems of sovereignty, terrain, local-isms, and supply can somehow be solved in an efficient or equitable manner. The very existence of such a mobile force owned by one particular state would be such a strong deterrent and such a prominent threat that it is unlikely the security concerns it creates for other states could be quickly or easily dealt with.

What about economic incentives? While the pursuit of profit is indeed strong, economic activity thrives in systems with greater disparities between the participants rather than few. There is reason to believe planetary unification could be both beneficial and harmful to global economic activity, and economists would therefore be relatively uninterested in a politically unified planet at first blush. This is primarily because the character of such a global market would be so unpredictable as to defy calculation, and from a traditional economist's perspective seems silly to even discuss. Since markets are driven by access to commodities and services which cannot be found locally, a global hegemony could conceivably be beneficial by easing trade—but it would not provide enough of an impetus to politically unify. After all, a global market already exists, and a global market with a universal political foundation would only serve to enhance or make easier current global commerce. In that case, what would be the economic point of politically unifying? From a market's perspective, the globe is *already* unified to the extent it could understand such a concept.

Nevertheless, looking at the potential for planetary political unification is useful. From a military perspective, a politically unified Earth would be able to harness and impressive fighting force, assuming the tricky issues of unity of command are dealt with equitably. Briefly discussed below are three potential methods of political planetary unification which aid our discussions about space warfare.

Potential Method 1: Security-Based Nation State Unification

It is conceivable at some point in in the future that, in order to adequately secure resources and protect status quo relationships, nation-states could permanently politically unify to better enhance their mutual security. Some states have grown politically close already based on mutual economic and security concerns, but these examples usually come in the form of temporary alliances or coalitions, and not true permanent federations. Usually, once the conflicts which threaten these states end, so does the alliance or coalition in question. One example that likely comes to mind when thinking about individual states politically unifying on a permanent basis is probably the European Union (EU). EU members carry a responsibility to defend the individual states should one come under attack, but at the same time each member state retains their sovereign right to make political alliances and untrammeled defense arrangements with states outside the EU.[23] The arrangement is at once bold and complicated, but exists successfully mostly due to its two main economic arrangements: an EU-wide customs union, and an EU-exclusive market to which members have preferential access and to which outsiders must negotiate for entry. These two characteristics make membership in the EU a profitable venture, and the security responsibilities contribute to the agreement's survival.

An arrangement made purely for security reasons, however, could best be represented by the North Atlantic Treaty Organization (NATO). Developed during the Cold War, NATO's almost exclusive mission to provide deterrence and counterbalance to the former Soviet Union and its satellite nations continues to this day in the form of deterring Russian ambitions. NATO members participate based on security pledges and minimum defense spending treaty obligations, and give up no sovereignty.

It is conceivable a future NATO-like organization could mature into a region-wide, then planetary-wide, sovereign entity, which not only holds global police authority but also some form of military authority. For the moment, however, a regional-based political entity has natural limits which bind such pacts culturally and ideologically. One of these limits is geography; technology has still not advanced to the point it can completely overcome the vast distances and oceans which prevent rapid cooperation between all NATO members. Another limit is NATO's need to find a foe to qualify its existence. The fact that NATO today is unified against a common enemy reveals its *raison d'etre* is at once both strength-based and boundary-focused. In other words, while NATO members are unified via a common threat and similar culture, they are also *bound* by these restrictions. NATO cannot realistically accept members too distant from their geographical limits (i.e. the North Atlantic), and also cannot introduce member states with neither interest nor stake in standing against Russia. NATO's mission would have to expand dramatically, and therefore change its purpose, in order to increase its membership beyond these boundaries.

A major factor which makes security-based nation state unification unlikely is the fact that such agreements exist solely to counter an already-strong politico-military threat, which implies their adversary is similarly unified or empowered to stand against said alliance. These arrangements also tend to drive their rivals to develop their own security alliances designed to counterbalance them, which defeats the purpose of unification. NATO, while a wise security move, stimulated the Soviet Union into forcing upon its satellite states the Warsaw Pact, a security alliance opposed to NATO. While the Warsaw Pact was a security alliance forged chiefly through coercion, it was a security alliance nonetheless. This makes security alliances a poor route to planetary unification, though theoretically possible only if one security alliance gets so powerful other nations have no choice but to join them. This of course begins to resemble imperium, and not voluntary unification. In this case, the last holdouts are likely to be the strongest nations who are most unwilling to submit their sovereignty to a hair-brained security alliance asking to hand over its military prowess. Security alliances do not bode well for blazing a trail towards peaceful planetary unification.

Potential Method 2: Response to Imminent Extraterrestrial Threat

This method is more of an emergency measure than a lasting political unification method, as it is essentially forces states to pool security resources in response to an external threat. In this case, the threat is a non-human competitor attempting to use coercion or force to adjust humanity's behavior. Many bickering states suddenly responding to an immediate security threat is neither new, nor are these ad-hoc coalitions ever permanent. Essentially, this method prescribes an urgent and demanding need for humanity to unite and face an extraterrestrial threat to ensure the survival of the species in a format which temporarily puts down our political concerns and focuses on an urgent security problem which affects our very existence. Popular in science fiction, such struggles are usually characterized by desperate technological disparity and dystopian societies clawing their way to some kind of parity against nameless extraterrestrial invaders. In one of the first and certainly most famous stories of this kind, H.G. Wells' *The War of the Worlds*, Wells points out the most likely outcome—the invaders win.[24]

While romantic and certainly exciting, there are several problems with this unification method. First, an extraterrestrial threat as clear as those which appear in films is probably not as likely as a distant or insidious threat. There are two reasons for this. The first reason is a sentient race attempting to threaten the very survival of our species, unless possessing an overwhelming preponderance of untouchable force, will be obliged to make initial first contact with us in order to exploit our inherent political weaknesses and divisions as they exist at the time of the beginning of the extraterrestrial enemy's operation. A shrewd attempt to manipulate the current political structure on Earth by enticing one or many political entities to the invaders' political side in exchange for assistance, technology, or the promise of victory and political survival could easily trick a nation or set of nations into neutralizing its political rivals on Earth, only to be betrayed easily and swiftly later by the invaders, either by guile or force. Such a method wouldn't even require a shot. The second reason is an extraterrestrial threat, if truly intent on dominating our world, would likely pursue a clandestine program first whenever possible. It is cheaper and easier to sow doubt and encourage disbelief about their existence to gullible humans, especially if the invaders are able to alter planetary media and communication sources through subterfuge and espionage. A slow and gradual *fait accompli* slows Earth-based military action, and allows the extraterrestrial adversary to strike when we are most divided, thus increasing the chances of a successful invasion. Finally, an extraterrestrial threat does not necessarily involve a planetary invasion of Earth. Destroying surface installations and life from a standoff distance, potentially even through chemical alteration of our atmosphere in attempts to exterminate our species, are all preferable to a risky, foolhardy and difficult planetary military operation. If this kind of threat ever materializes, most human victims will likely never see the face of their killers.

The second problem with this planetary unification method is our history clearly reveals that once a mutual enemy has been removed by a set of political entities, those entities return quickly to previous squabbles or selfish political concerns. To be fair, this seemingly ungrateful phenomenon is more due to our nature as primates and to the political realities we have learned about ourselves over time than anything else. Trust is simply something that does not always come from mutual success on the battlefield. One famous example of this phenomenon is the inability of Arab states to capitalize on their contribution to Allied victory in Arabia and Egypt during World War I to form a solid unified political foundation which could later properly challenge British rule. Although the British could likely not have been as successful during their campaigns in the Turkish-held territories in Egypt and the Middle East without the help of local Arab irregulars, the various Arab tribes who assisted the British quickly reprioritized their political realities after victory, especially Egypt.[25] Egyptian nationals in March 1919, who only months before looked upon the British Army as their liberators from Turkish oppression, erupted into nationalist riots when they realized British rule was not going to disappear. Rather than using their newfound partnerships with Arab tribes to collectively drive out the British, Egypt happily pursued unilateral independence and returned to its ancestral disputes with Arab tribes.[26] This left Western powers deeply involved in the region until the end of World War II.

In truth, alliances for the purpose of defeating a common enemy are often more precarious than they appear from outside. In the case where longstanding rivals unite for this purpose, there is a virtual guarantee they will return to eyeing each other suspiciously after the urgent security problem is solved. Oftentimes allies will attempt to weaken or discredit each other at every possible opportunity if one or both sides smell victory. One need look no farther than the Nationalist and Communist Chinese forces near the end of World War II. While both forces were ostensibly united in purpose to push out the Japanese Imperial Army, all the while they eyed each other with suspicion and maneuvered carefully to prepare for the day they would face each other in battle again.[27] Civil war between the Nationalist and Communist forces indeed broke out in 1945; after Japan's surrender, both sides scrambled to occupy areas formerly held by Japan and to assert their brand of authority on a divided China.[28] This is clearly no way to unify a planet.

Potential Method 3: Regionalization Due to Inter-Regional Competition

Of our three distasteful choices for planetary unification, this method is the least distasteful, but bitter nonetheless. As technology and economic pressures continue to grow, and continue to be fed by demographics and national security threats, it is conceivable that regional economic pacts could lead their way to full-fledged defense and shared-sovereignty pacts designed to ensure survival in a world which has quickly become too complicated to manage as one nation alone. As the world continues to shrink, mostly due to technological advances, states may be forced to surrender their previous authorities and sovereignties to a greater, more responsive unified command structure in order to decrease response time to incidents, pool resources, share technology, bolster standing military forces in the face of declining birthrates, fight international crime, secure borders, and present a more threatening face to potential adversaries.

Note the operative words here: *may be forced to.* It is unlikely comfortable sovereign states would voluntarily unite in this fashion without existential threats. The cultural costs to such a unification would obviously be high as well. Consider, for instance, a political unification of Canada and the United States. While culturally the two nations are very similar, there are enough differences to cause trouble with hard-liners on each side, and there are many who would decry the abrogation of several hundred years of history for a new-fangled and untried security arrangement. Disagreements would range from location of the new capital, to the shape and size of the armed forces, to regional resource prioritization; the list goes on. If possible, such an arrangement would prove militarily formidable indeed: the population reservoir alone, along with uniting the entire northern half of the Western Hemisphere under similar values and a committed defense, would terrify potential adversaries and act as a catalyst for culturally similar states to copy the arrangement as they sought protection in such a world.

Indeed, it is conceivable that the first such regional unification would see a landslide (at reckless speed) response to similarly unify in other regions as well. When one considers international relations dictum that human political systems tend naturally to try and balance threatening powers, a unification "domino effect" could conceivably occur as states elsewhere sought to protect themselves against the suddenly threatening new U.S.-Canadian state.[29] In a self-help system, states would have no recourse but to make themselves strong enough to potentially fight against this new monstrosity, and it is doubtful they could do so without the help of other states. We can expect nations which have existed culturally for a longer amount of time, like Asian and African cultures, to predictably unify more around cultural and historical grounds rather than geographic and ideological. In the end, such political activity would produce a world not of 200-plus countries, but one of five or six extremely messy ones.

Such arrangements are clearly rife with trouble, if they are even possible. This is no surprise to a reader who has conducted foreign diplomacy; the consultations alone required to form a multi-national politically unified sovereign state are petrifying. Any nation attempting such an arrangement would have to understand the problems which will never go away and which must be monitored forever: political discord, nationalist yearnings, haves and have-nots created by rich countries uniting with poor ones (and vice versa), resentment of traditional authorities thrown from power, outbreaks of crime and terrorism, government inefficiencies, and economic inequity. Risk of war will also increase, as the existence of fewer states will promote greater suspicion between the ones who are left. With fewer political entities available to balance out state-to-state rancor, risk of conflict will certainly increase. But if it can be managed, and managed peacefully, such an arrangement has a chance to lead to planetary unification. Once the unifying is done and a suitable balance of power is found or forged here on Earth, better and more sophisticated cooperation could begin in space.

While as of this writing these pan-continental sovereignties are inconceivable, the only real obstacle to making them possible is technological development. In our U.S.-Canada unification example, a citizen from northern Manitoba must be expected to be able to participate in an election, on the same day, as someone in southern Florida; they must both send their taxes to the same tax authority; they must both have police protection available within the same approximate response time; and the armed forces must be able to deploy to either location in a timely manner to face external threats. In many ways, these tasks are already being done; the biggest differences, therefore, are political. The ultra-fast transportation and efficient security required of this arrangement are not yet available, but given time and technology, there are no conceivable limits preventing effective administration of a domain of any size.

This chapter discussed some human ideological concerns as we approach deep space warfare. While the predictions here are by no means guaranteed, they are designed to get military leaders and decision-makers thinking about the consequences of human political thought, ideology, and technology on things we may encounter as we continue to step slowly towards deep space weaponization and war in the newest domain.

4

Space Dominance

> "Months of boredom punctuated by moments of terror": such is a description of life in the Navy which a naval lieutenant quotes as exactly fitting the facts.
>
> —Edward Arthur Burroughs

Since 1982, the U.S. Air Force, the world's pre-eminent air and space power, has experimented with just what "space doctrine," or fundamental principles by which a military force must operate to obtain dominance in space, should look like.[1] Since its formal promulgation in 1982, space doctrine has been through a lot. Its first form was fashioned in the image of terrestrial air power prerogatives, then rescinded, then rewritten and re-promulgated, then ignored, and—presumably to allow for the greatest possible understanding—re-fashioned once again in a shape resembling terrestrial air power doctrine. This refashioning, it is hoped, will assist the mostly-disinterested USAF officer corps to better understand space's complexity and consequence.

Why all the false starts? Several reasons have prevented progress in clarifying and codifying space doctrine, without which efforts to seize space dominance are difficult to organize. First, the U.S. Air Force suffers from unwritten cultural proclivities which tend to discourage doctrine in general as a prominent part of warfighting. This engenders hesitation in crafting space-oriented doctrine and policies. Second, two major historical points of inertia tend to discourage advances in the space realm. The first point is the U.S. agreement with its old foe the Soviet Union to refrain from weaponizing space, which at the time was a real and measured decision to prevent a runaway arms race.[2] While this policy kept the peace, it also discouraged active weaponization research and thought about dominating space from a conventional point of view. The second point is the natural inclination of any Great Power at the top of its game, in this case the United States as the world's only superpower, to refrain from innovating in a realm which it considers itself dominant.

The second, and most consequential, major reason why U.S. space dominance policy has never measurably got off the ground is the lack of a noteworthy adversary in space to incentivize a need to innovate.[3] During the Cold War, space dominance was assured as far as either country was concerned; both the United States and the Soviet Union could launch their intercontinental ballistic missiles into space whenever they chose, with no obstruction, with little warning, and with little to no effective defensive measures available. Space, therefore, for most of its history has been a geopolitical battleground primarily safeguarding terrestrial security concerns, and secondly for exploration and bragging rights.

Beyond space acting as an occasional and inconvenient battleground, for much of its history space has been a convenient place to launch better and more secure telecommunications equipment, a place to launch tremendous stellar observatories such as the Hubble Space Telescope, and a medium which provides a critically important but expensive scientific laboratory.

This has all changed with the sudden transformation of space by three major powers—Russia, China, and the United States—into "the new high ground." As those nations become more and more dependent on technology and telecommunications to secure their national wealth and prestige, space has become more and more attractive as a place to secure those interests. Moreover, several major treaties have attempted to equitably balance the ratio of satellites and orbital objects each country is allotted; these and other treaties have succeeded, to some degree, in addressing some problems with conscious concern over their potential to spark new conflict on Earth.[4] Unfortunately, no treaty will prevent a determined political power from pursuing enough control over space to satisfy their perceived security requirements. When one political unit achieves a preponderance of this control and power, it is called "space dominance."

To be clear, this book addresses "space dominance" not in the terrestrial-focused, air domain-inspired definition of space as simply another place to extend terrestrial battle higher. Space is not *only* a place to "seize the initiative" and "maintain control"; truly, these are military objectives for *any* domain, terrestrial, naval, airborne, or other. Space is also a broad-reaching mandatory objective to obtain for any space-faring military fighting against another space-faring military. "Space dominance," then, refers to a preponderance of space weapons, vessels, and other space-going instruments of war which enables a localized predominance of military power. This does not have to be only in quantity, but in quality, used when needed to maximize coercive power against a spacefaring adversary. It also means a preponderance of control and the ability of the party with space dominance to deploy forces at the time and place of their choosing.

To two spacefaring civilizations, space dominance ought to be the first and most urgent objective prior to engaging in battle. With space dominance, each planet and each system begin to resemble ports and seas, respectively. Recall our earlier discussions about "all-or-nothing" space combat. Dominance provides extra reassurances and greater chances for a particular force to achieve a more complete victory during space combat. While total destruction of an enemy force is not always the right strategic goal, it is very nearly always tactically sound.

Achieving space dominance can be best accomplished by the navy with the greatest preponderance of power.[5] What good are splendid terrestrial air forces, a fine terrestrial maritime force, and a sophisticated army when all three can be locked onto a planet and bombarded from orbit by a space blockade? It stands to reason that a stronger space force, predominantly in space naval superiority and strength, is the key to space dominance. Such forces could imprison whole enemy hosts on their planets, in their systems, or sufficiently separate enemy forces and territory from others. A strong naval presence forces the enemy to whittle away his time, expend resources, look elsewhere for less trained or ready forces, expend precious fuel and time seeking holes through which to bolt through a superior navy's line. In short, achieving space dominance seizes the operational tempo of any space campaign.

In space dominance, therefore, encircling enemy forces and territory must be a primary objective. As Hannibal and Bulow have shown us, nothing beats a good double envelopment; not only are all forces within smashed and rendered useless, with the survivors at the mercy of the conqueror, but also the blow to national morale from suffering a double-envelopment is second to none. In the case of Sedan in 1871, the blow was unrecoverable and let to French defeat. Only the Romans were able to recover from their great encirclement defeat at Cannae and go on to win the conflict, a fact underscored by the reality that Rome itself remained free to continue to raise legions and politically able to fight. Paris was not so lucky.[6]

There are those that would say destruction of the enemy, as Nelson once did, is the best policy.[7] Nelson's endorsement of this policy is particularly strong given his legendary status and remarkable success on the waves. But Nelson's pursuit of tactical dominance does not necessarily translate perfectly to space combat. In space, a war will more resemble a game of *Go*[8] than it will a game of chess. The former aims to encircle an opponent and deny him his strategic maneuver; the latter aims to deprive the opponent of his pieces completely and slaughter his ruler. In a world where battlefield losses could not be replaced during combat, where technological innovation would freeze to a standstill, and where no new pieces could be added to the board after the game begins, then chess would be the logical choice. However, with the span of time we expect an interstellar war to encompass, when considering the vast distances forces will be required to traverse, and the sheer amount of ranging materiel and planets potentially available as a battlefield, it is clear any successful wartime strategy is not necessarily on the side with the most tactical victories. As we shall see, the distances and time associated with space warfare will prevent tactical triumphs from significantly altering a strategic outcome if they are not capitalized upon immediately and in combination with other forces. Encirclement and denial will be much more important during a space war than relatively short-term force destruction.

Like any doctrine, space dominance is not without its share of problems. Some of those are illustrated here.

The Trouble with Space: The "Never Ready" Blues

Military planners are never really ready for campaign. While they come in many varieties, planners are generally gloomy and pessimistic. They naturally gravitate towards the worst-case scenarios and never cease their hand-wringing about the reliability of supplies, consistency of transportation, the weather, or even the reliability of allies.[9] Indeed, a military planner who claims to have everything ready and sure of himself should never be trusted. Custer famously thought he was ready, even over-prepared, for what awaited him at the Little Bighorn. French Marshal LeBoeuf famously claimed the Imperial French Army was ready for combat with the Prussians "down to the last gaiter button" prior to the Franco-Prussian War; that army later found itself surrounded and starved in Sedan, in no small part due to the lack of having plans drawn up for a general offensive.[10]

It is commonplace in military planning that the plans themselves regularly take longer to formulate than the campaign which draws upon them. This makes sense; after all, given the stakes of any military plan, it is worth an extra look or two. We can safely expect the same for spacefaring campaigns, especially those involving movements of troops or large amounts of materiel. Space military planners, however, will face a great deal more adversity than traditional terrestrial planners.

One challenge is time. As mentioned in an earlier chapter, preparations for any space military adventure will likely begin years in advance. This is due to the time required to muster forces (if using forces dependent on human soldiers), gather supplies, construct vessels, await the necessary change of stellar phenomena or space weather, build automated forces, train assault forces and crew, re-tool economies, transport supplies from outposts to homeworld and vice versa, and obtain the necessary—and lasting—political backing to approve the venture. These efforts will be compounded and complicated by rapid military technological progress; in other words, military planners will likely be confronted with *too many* options from which to choose specialized equipment and technological solutions rather than too few.

A fictional example is illustrative. Previously we discussed how assault forces must be specifically tailored to their target planet to maximize their chances of success. In the distant future, planning for such an interstellar planetary assault would begin several years prior to mission launch, with the technology available at the time; personnel would begin to muster, supplies stockpiled, transport spacecraft keels laid down. However, technological advancement will continue regardless of military needs, and military preparation on the scale required for interstellar assault could not likely be totally concealed from the general public. The streams of equipment requests which would presumably occur from the government would not escape notice. The ever-present military contractor, eager to earn money and political influence during conflict by using the most advanced (and expensive) technological capabilities their companies can muster, would no doubt pour much of their military contract earnings into research and development to offer newer, better equipment to solution-hungry military leaders. New assault vehicles which more suit the target's terrain would suddenly become available; several new types of body armor advertised to keep troops safer would appear; ammunition which penetrates the foreign atmosphere faster and truer would suddenly leap out of several laboratories; longer-lasting rations, better solvents to treat corrosion from war machines afflicted by the alien atmosphere, more efficient engines to propel assault force starships—a looming planetary invasion could spawn such offers during the multi-year preparations made by military forces. Companies eager for a slice of the suddenly-higher military budget will only serve to complicate plans as the government in its eagerness to win accepts contract after contract.

As we know from history and current military operations, many of these contractor products promise much but deliver little; and there will always be limited time to test them before the invasion must get underway. As one new innovation or another in turn is revealed to be a dud during military testing, it becomes one more wasted military investment and one less planned capability which military planners now have—money and energy which is gone forever. The pressure on planners to deliver the absolute best equipment and supplies prior to an undertaking of such a magnitude as an interstellar operation would be enormous; even the slightest chance to increase combat success may be taken with these stakes, with an entire species quivering with the nervous energy of war.

The solution to this potential future conundrum is easy, while at the same time makes military leaders uncomfortable: select the desired technology and capability which has a reasonable chance of success, and stick with it, while minimizing expensive upgrades and unproven "pie in the sky" capabilities. A strong argument can be made that once the assault force begins training, unless catastrophic problems have been discovered with their equipment, their equipment should not be altered unless absolutely necessary. A soldier going to war with a dependable weapon which he understands well is much better prepared for battle than an experienced soldier with a new and untested weapon thrust upon him. This idea flows throughout all aspects of military operations—from ship pilots to logisticians, from maintenance personnel to cooks—humans demand familiarity, which begins first from being comfortable with a checklist, piece of equipment, weapon, or procedure. Without adherence to "what works," space borne forces are doomed to "chase technology" and will always be prisoners of the next best piece of gear without taking the time to appreciate or extract all possible value out of the gear they already have. This is, sadly, a lesson which terrestrial armed forces have had to learn many times over.

Military planners will have to come to accept the "never ready blues" as an essential component of planning any space operation. On the bright side, though, constant fear of being unready or underprepared will likely trend towards overpreparation, which always has the benefit of increasing safety margins. The trick will be when to know where to draw the line with military preparation.

The Military Need for Settlements

Colonialism is, deservedly, a hated idea in the 21st century. As a concept which once was unabashedly referred to by Europeans as "the white man's burden," the notion of taking another's land simply to improve the natives, civilize them, exploit them, or some combination of the above, is now widely regarded to be offensive and destructive. The late 20th and early 21st century have clearly demonstrated that previously hatched colonial chickens are still coming home to roost; former colonies find themselves trapped in an endless cycle of poor leadership, autocracy, and disadvantaged economic competition which has doomed their populations to poverty and obscurity without end. On Earth, colonialism, it seems, is a gift which keeps on giving.

Extraterrestrial colonies, better known as settlements, are a different matter entirely. In their case, settlements represent a group of humans, supported by the state, who venture voluntarily into the great stellar expanse to seek better lives and expand humanity's home. While these settlements mean different things to different people, to military planners settlements mean additional sources of state's extended power, greater security obligations, and potential sources of military materiel. Some of the complexity involved in interstellar settlements has been explored by science fiction, but their conclusions are immaterial to this book, which only concerns itself with extraterrestrial human settlements from a military perspective. The military need for settlements becomes clear in simple, cold, mathematical terms.

To be militarily competitive in any interstellar region, quantity often has a quality all its own. Given what has already been discussed about the difficulties associated with fielding useful spaceborne military forces, the economic resources required to support the tremendous efforts needed to project interstellar power and conduct interstellar operations require extraterrestrial resources. A very simply mathematical calculation demonstrates why. In this example, our future planet of nine billion people finds itself in conflict with a group of humans on another planet, or another sentient civilization, who happens to have a similar sized planet which is similarly populated. To make war on these beings, it stands to reason both governments can only mobilize a certain percentage of their populations, and will do so at the highest rate possible since their very existences are at stake. This mobilized percentage will of course depend on things like domestic attitudes and political will; but by and large the populations and their available forces in this example are evenly matched in population resources. Thus, the combatants have a few choices: conduct drastic military operations to even the odds, like employing weapons of mass destruction, or find a way to mobilize *more* than their adversary. In any case, the more a force depends on weapons of mass destruction, the more they stand to lose for themselves upon ultimate victory; a bombed-out and radioactive planet is much less valuable than one with a viable atmosphere and untouched topography. Another choice, conventional conflict, is a grueling prospect. Any force-on-force conflict where both sides have approximately the same strength typically leads to grinding slugfests where both sides suffer greatly.

Neither of these choices are favorable. There would be little prize left in a strategy of obliterating the enemy planet through bombardment by destructive weapons, or through economy-crushing force-on-force warfare between similarly equipped populations and economies. This means a war between equal populations in an interstellar conflict favors perpetual and brutal conflict, and reduces the incentive for peaceful conflict resolution the longer a fight goes on. Such prolonged fighting also tends to raise the stakes of victory and overcharge emotional involvement, and will likely end in the eventual destruction of the planet that loses the upper hand, no matter what the original intentions were. This end will doubtless come in horrific ways, such as nuclear or chemical bombardment on a planet-wide scale, and will lead to nowhere but endless hate and war.

But imagine the same scenario once more, this time with one side possessing an extraterrestrial settlement. Suddenly, the game dynamics are completely different. The side with the settlement is free to use the settlement as a manufacturing and military base; to continuously threaten the enemy from another azimuth of attack with garrisoned forces, like a medieval castle; to act as a lifeboat in case its homeworld requires evacuation during a losing struggle. In all cases, the side with the settlement possesses clear advantages. Fresh forces and supplies from the settlement are free to reinforce an attacking force elsewhere or to deploy on a gambit to take pressure of their homeworld under siege. They could pounce upon enemy assailants from the rear, or to attack the now-defenseless enemy homeworld, an enemy who has no settlement and whose forces are occupied elsewhere. This leaves the side with the single planet—its homeworld—to risk everything when sallying its forces to attack its foe. During such an assault, the side with only one planet can only hope for a quick victory, or that their foe's settlement or settlements are none the wiser.

Once a belligerent possesses settlements, interstellar warfare with the same approximate parity between powers no longer risks mutually assured destruction. The force with no settlements stands to be defeated permanently if they lose, while the force with settlements has a place where its species can survive and recover if their homeworld falls. In short, settlements act like castles in the ages before gunpowder; while susceptible to siege and conquest, on the campaign map they cannot be ignored, lest their garrisons sally into the enemy's flank or rear. Even better, planetary settlements, assuming they are self-sufficient, provide an even stronger hold than a classic fortress (which requires constant resupply), and thus an even greater threat to an interstellar foe.

Settlements are necessary, and the more the merrier, for other military reasons. One such reason is the clear military benefits of greater population and economic growth, which eventually translates to greater military power. This is due not only to the greater number of people who could realistically serve a military operation in some capacity, but also due to a greater industrial and agricultural output propelled by a larger population, not to mention innovation and leadership sources the extra people would provide. To maximize a settlement's value, each planetary settlement must also be ideologically loyal to the homeworld to the maximum extent possible. A settlement must be counted upon not to crack under foreign military pressure or subterfuge, nor act as ticking time bombs of dissent which the homeworld must always worry about. If we keep with our North American British colonies example from above, planetary settlements must never be allowed to sink into salutary neglect, lest they completely lose their military advantage to the homeworld.

When deciding when and where to place a settlement, nature makes much of the decision for us, or at least will initially. Planets which could host the human species are obviously of greatest interest, which are found in the famous "goldilocks zone" in a stellar system.[11] This refers to an appropriate distance from a system's sun or suns where liquid water is regularly available in relatively high concentrations on a planet's surface. A planet too close to its sun(s) would be too hot for liquid water to remain constantly present without evaporating; too far, and water would be frozen and therefore unusable for biological life as we know it. The presence of liquid water also generally indicates potential for a nitrogen-oxygen atmosphere, another major requirement for settlement candidacy.

Doubtless our species will have very little choice in what planet would work for colonization prior to our first exoplanetary colonization attempt; as we are discovering, there are not that many Earth-like planets in the neighborhood, let alone close by. According to the open-source database of exoplanets discovered, the *Open Exoplanet Catalogue*, of the 3504 confirmed exoplanets thus far discovered, only 310 appear to be in the "goldilocks" habitable zone, more formally recognized as the Kepler Habitable Zone.[12] On average, according to current data only one in five sun-like stars contain an Earth-like planet whose orbit is firmly in the Kepler Zone which could allow liquid water to pool on its surface.[13]

While it is conceivable that humanity will one day have the technology to completely alter barren and inhospitable worlds to a more pleasing and livable condition, a process known as terraforming, that tremendous engineering project remains well out of our reach and relegated to science fiction for now. For the foreseeable future, we can only pursue what nature has prepared for us by chance.

The decision to settle an Earth-like exoplanet should be based as much as on cold mathematics as on our human desire for discovery and challenge. Settlements should be thought of in terms of a return on investment rather than a pure ideological adventure. If a group of, say, 25,000 human settlers

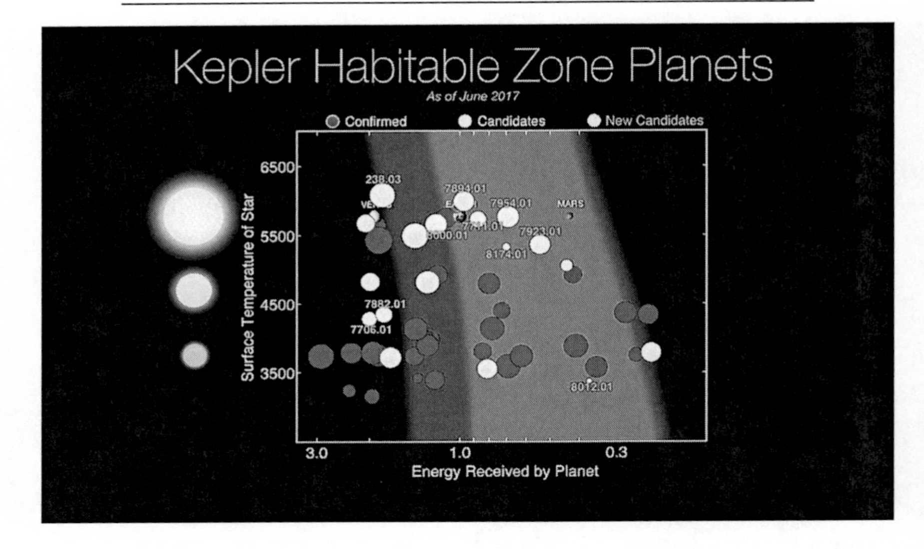

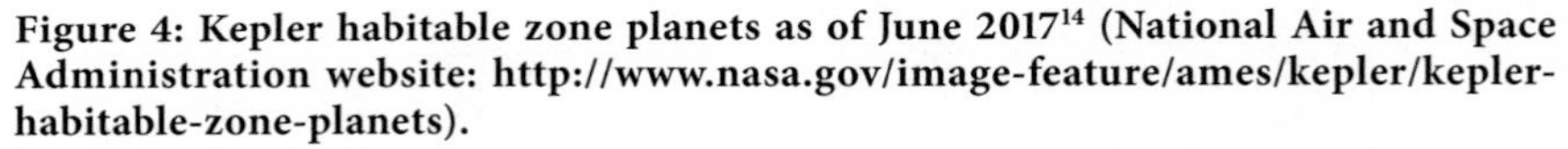
Figure 4: Kepler habitable zone planets as of June 2017[14] (National Air and Space Administration website: http://www.nasa.gov/image-feature/ames/kepler/kepler-habitable-zone-planets).

were to settle upon an Earth-like world in the not-too distant future, how long would it take for them to populate the useful parts of the planet, given the technology and medical care they would have? Assuming they remain politically unified and subject to Earth in terms of government and taxes, at which point would a settlement begin to benefit its home world in military terms vice acting as a drag on its economy with its constant need of support while it tames a new frontier? Clearly, a policy of continuous and relatively open settlement—that is, immigration—must be pursued if a settlement is to benefit its master during its founding generation. Massive incentives to settlers must also be used to get a settlement up and running as fast as possible, to be militarily viable as soon as possible.

Once an extraterrestrial settlement can stand on its own feet, the military benefits to the home world are tremendous. In warfare terms, the added population provided by a colony is absolutely necessary to assist a spacefaring military scenario requiring the most manpower of all scenarios—a planetary invasion capability. The math is clear: to assault an entire planet which is of equal or greater size than one's homeworld, the attacker will need a greater population that that of its homeworld alone in order to generate the extra engineers, builders, suppliers, businessmen, manufacturers, equipment, products, soldiers, vessels, and anything at all on the scale needed for a planetary invasion effort. Even in the case of entirely automated soldiers and battle fleets, the space, raw materials, agricultural output, and manufacturing scale required to build up the necessary forces for an offensive invasion would likely need at least two worlds to take on the project, similar to a car company having more than one manufacturing site to keep up with demand.

Indirect Military Benefits of Settlements

While this book concerns itself with military matters, it would be remiss if it failed to point out the military establishment of settlements would simply be the latest in a long line of military technology with dual-use applications for the civilian sector. First and foremost, settlements are the ticket out of mankind's economic imprisonment on Earth, with the Law of Conservation of Matter as its chief warden. While this law has many applications in fluid dynamics, chemistry, and physics, here it concerns itself with its effect concerning closed systems. The law states that in any closed system, the amount of matter (or mass) is constant. Regardless of how molecules within a system arrange themselves, the sum total of all matter available for this rearrangement never changes. Planets like Earth are not technically closed systems; the occasional asteroid strike adds minuscule amounts of matter to Earth, and there is a great deal of energy transfer into and out of our atmosphere. However, on a larger scale mankind will always fight with itself and the mass on the planet to find the best ways to rearrange Earth's matter which can better support itself and its continually-growing population. In other words, mankind's need to feed, clothe, arm, and care for itself acts as a natural drag upon and impediment to activities beyond our planet.

A spaceborne settlement, therefore, simply enlarges our current one-planet system into *two* planets, thereby expanding the principal size of the system under which the warden must watch. By expanding this system using robust trade between our homeworld and settlement, the mutual material problems can be better alleviated and less energy needs to be put into methods which are designed to rearrange Earth-only matter to solve the same problems. Economic concepts like these will be discussed in greater detail in Chapter 6.

Planetary Systems as the Key to a Strategic Stronghold

Every form of warfare possesses critical objectives which are keys to winning a campaign. In military parlance, objectives of paramount interest which cause victory to precipitate more readily for one side or the other is called a Center of Gravity (COG). While these are often discussed at the strategic level, they exist tactically as well. Traditionally, terrestrial battles concern themselves with tactically advantageous terrain, especially high points of elevation, to view the enemy, direct artillery, and obtain an advantage in land-based slug fests. Sailing fleets in antiquity sought the weather gauge—the direction in which the wind blows—which was used to secure tactical mobility over their opponents. Modern fleets seek key maritime geographical features to force enemy fleets into bottlenecks and restrict their mobility and sailing routes, block their access to friendly ports, and simultaneously pressure their international trade, supply chains, and fuel reserves. These tactics have clear applications for future deep space warfare.

Previously we have discussed what space is—mostly vast and empty as far as the military is concerned, with the tiny fraction of matter in the universe mostly fleeing the darkness via gravity's pull towards stars and other objects of larger relative mass. Thus, near these objects of mass—especially stars—are nearly everything humanity will find useful in the cosmos: the stars themselves, with their seemingly limitless fusion factories as energy resources; planetoids and rocky bodies which no doubt conceal vast mineral deposits and yet-to-be discovered elements; every kind of planet, colonizable or not, which even if it does not support our kind of life will surely lead to tremendous and valuable scientific discoveries; and many more as yet unknown treasures our species will surely come to value.

It is this lonely fact—that star systems contain nearly everything we find materially useful in space—that make them our future strategic strongholds. Further, because star systems contain planets and resources, they will also act as the future harbors of deep space vessels, places to find dependable gravity (to act as location anchors, such as a planetary orbit) and energy sources (especially solar) to refresh future space farers and provide valuable locations for way stations and military supply points. Star systems also provide concealment: they host multiple layers of electromagnetic activity to confuse sensors, contain massive objects to hide behind, and are affected by electromagnetic forces which can be used for tactical purposes, as we shall see below.

Deliberate Targeting: A Challenge to Prioritization

Any plan to strike enemy systems with the purposes of seizing or neutralizing parts of them immediately reveals multiple complications. The first, and perhaps most important, idea is: what to strike first? Expeditionary forces will have limited resources, fuel, and munitions, further complicating operations further afield (and incidentally making the case for certain types of weapons, like directed energy, over others, like explosives). Finding the right target which delivers the most impact on the adversary's war effort while simultaneously maximizing one's own advantage is a perennial challenge for every strategist, but not necessarily every statesman. This means that, as always, the decision to attack a particular target and what priorities should be assigned to targets should be made by military authorities and then approved by civilian ones. Military leaders and strategists develop operational plans based on an overall military campaign goal. This could include eliminating the conventional forces of an enemy, the taking of a particular territory, halting an erupting conflict in a region, liberating a nation under military rule, forcing a nation's surrender, and so on.

Above all, targets are selected first and foremost for their strategic effect. Destroying nearly any target has an effect; but whether or not that effect produces results which aid a political goal is another matter. The U.S. assault on Iraq in 1991 designed to remove its army from Kuwait, known as Operation

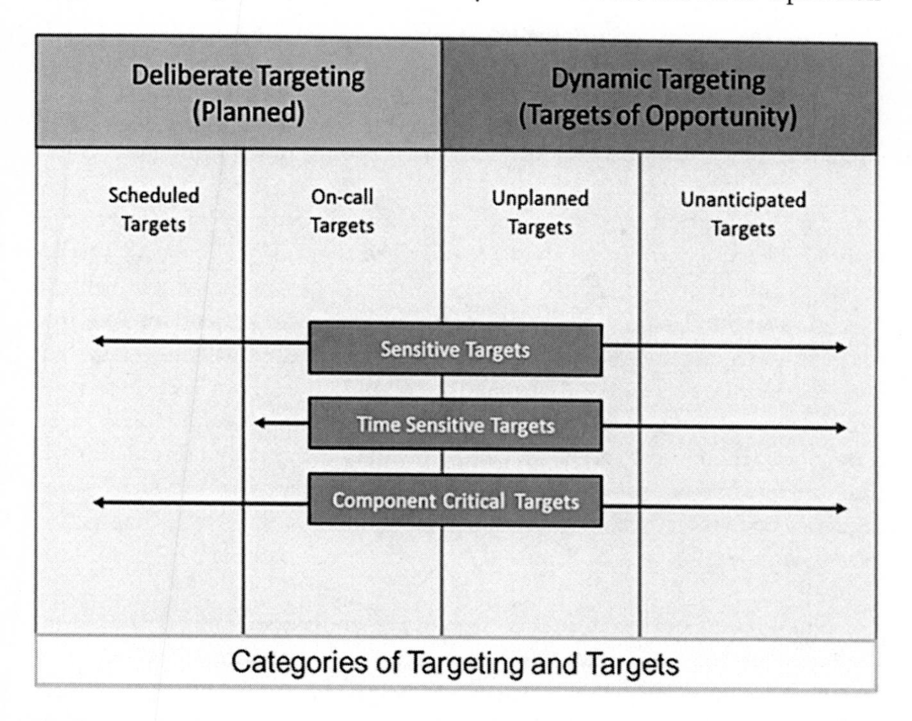

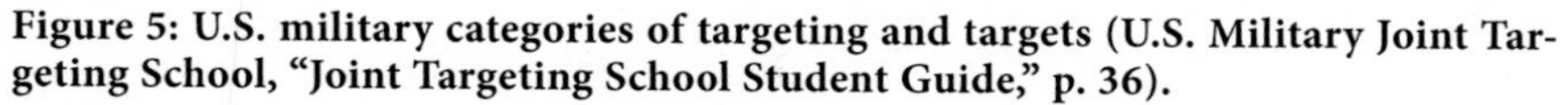

Figure 5: U.S. military categories of targeting and targets (U.S. Military Joint Targeting School, "Joint Targeting School Student Guide," p. 36).

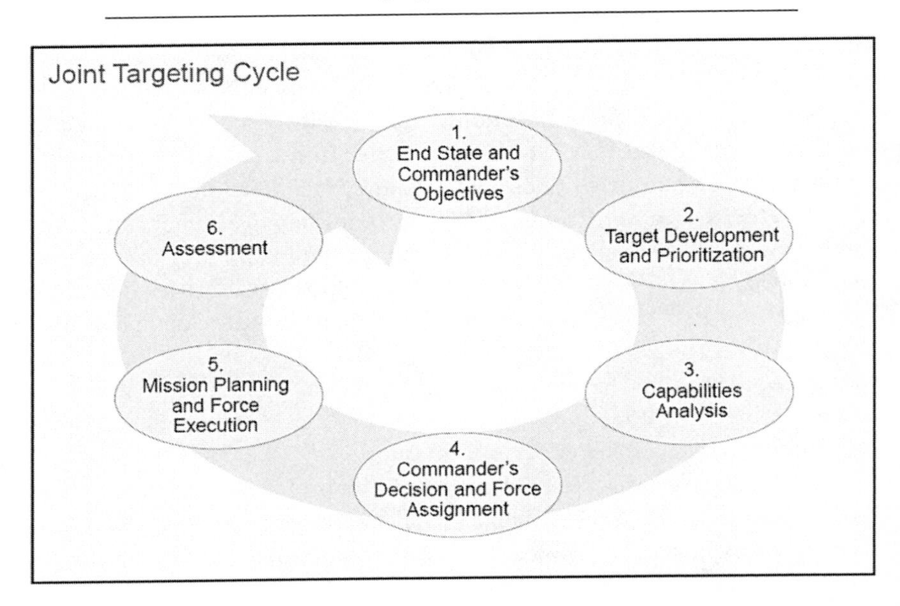

Figure 6: U.S. military joint targeting cycle (U.S. Military Joint Publication 3–0, *Joint Operations*, p. II-4).

Desert Storm, provides an excellent case study on target selection.[15] Scholar and USAF officer Colonel Edward C. Mann III points out planning can reflect three different options for targeting.

The first option is to attack targets in a series with no regard for their individual value, where strategic results are the sum total of all targets attacked and destroyed.[16] This means each target has an equivalent value; a target is a target. Planners need only focus on the level of destruction required for each particular target, then select the best weapon and delivery system available to guarantee a good probability of kill.[17] In this method, there is no need for prioritizing targets since all targets are equal in value; forces need only attack the easiest target first until the target list is exhausted. This is often referred to as "serial targeting," or destroying targets in a one-by-one sequence until there are no more. In this version, the series itself is haphazard because the targets are not required to be destroyed in a particular order or priority.

The second option available to planners is to assess targets based on value, then determine the strategic results when they are destroyed based on a weighted scale.[18] This means targets are assigned a value which then drives a priority, with the more valuable targets ostensibly being destroyed sooner to make an overall greater impact on the enemy for a longer period of time.

Theoretically, this option is supposed to bring victory faster since the enemy is deprived of more important items sooner rather than later.[19] Nevertheless, this option still falls under "serial targeting" if the targets are destroyed sequentially rather than simultaneously.

The third option, and the targeting plan developed by USAF Colonel John Warden and his planning team, later called "Instant Thunder," can be termed exponential strategic impact targeting.[20] By attacking multiple specific targets simultaneously against vital centers of specific enemy systems, planners seek a systemic "catastrophic failure" simultaneously rather than waiting for progressive system failures from serial attacks to take effect.[21] An attack such as this, if successful, has the effect of inducing what planners call a strategic paralysis. Like a boxer hit by a flurry of punches, under strategic impact targeting the boxer "loses control of his vision, brain, central nervous system, and body."[22] Unlike the serial targeting of the first two options, this option calls for parallel targeting, or the simultaneous or near-simultaneous neutralization of multiple targets designed to induce strategic paralysis. This is just what happened in "Instant Thunder"; simultaneous strikes on Iraqi communications, power generation, and command and control facilities blinded Iraqi leadership to the situation as it developed on the battlefield and temporarily paralyzed Iraqi decision-making. Strikes thereafter induced a feeling of panic and hopelessness in Iraqi leadership and fielded forces.[23]

Wars like these prove to us proper targeting should produce a strategic effect which helps achieve a political objective. One such strategic effect is the aforementioned strategic paralysis; another is British strategist B.H. Liddell Hart's concept of a strategic dislocation.[24] This concept is slightly different than a strategic paralysis, but is just as useful for targeting. Hart calls a strategic dislocation a physical effect which upsets an enemy's dispositions, dislocates the distribution and organization of his forces, separates them, endangers their supplies, or menaces the routes by which the enemy can advance or retreat.[25] While Hart advocates for an "indirect approach" to achieve a strategic dislocation, Warden's plan was anything but indirect. By striking at the nerve centers of Iraq, U.S. forces were able to inflict strategic paralysis and dislocation upon Iraqi forces through parallel attacks.

During interstellar and interplanetary deep space warfare, targeting will be similar to terrestrial combat in that all targets should be destroyed to affect a strategic paralysis on the enemy. It differs, though, in the logistical challenges of scale, especially when a target set could include a sizeable area like a planetoid, continent, or even an entire planet. Weapons must be either carried the great distance required to reach the target, launched with improbably good accuracy from a safe and distant location, or manufactured at a staging area near the target. The weapon systems required to employ the chosen weapons, be they vessel or soldier, must also survive the voyage intact to the target. One can easily see the logistical challenges targeting presents, regardless of the tactical method chosen to execute an attack.

The United States military today provides several examples of the targeting process which could be emulated in future space conflict scenarios as long as space's restrictive characteristics are taken into account. While each military service must concern itself with its own targeting priorities, for example, land interdiction by the U.S. Army, for the purposes of this text U.S. Air Force Pamphlet 14–210, *USAF Intelligence Targeting Guide*, will suffice as a basic introduction. It, along with Air Force Instruction 14–117, *Air Force Targeting*, has the added benefit of being a well-polished and battle-proven document which can act as a solid foundation for building targeting procedures for space campaigns.

The following diagram represents a basic military targeting concept.[26]

The military targeting process is no more than a massive and constant cooperation between operations and intelligence personnel before, during, and after campaigns and combat operations.

All military targeting essentially comes down to weapons prioritization and target value. One aspect of targeting not often considered by non-military observers is the fact that weapons are often limited, which in modern warfare is a main driving factor to prioritizing the right targets. There are only so

Figure 7: One USAF targeting concept (Air Force Pamphlet 14–210, *USAF Intelligence Guide*, 1 February 98).

many airplanes which can only sortie so many times in one day; only so many cruise missiles on guided missile destroyers; only so many bullets in a rifle. In a space campaign, this shortage will be accentuated by the time and distance it takes for an offensive force to reach the target area, whereupon intelligence will likely have been revealed to have been wrong, the natural fog and friction of war to have concealed more targets than first thought, or the entire nature of the campaign could have changed. What could have started as a reconnaissance-in-force could have changed to, due to other battles or a worsening political situation in the intervening travel time, an all-out punitive campaign.

Finally, before targeting is finalized, targets must be selected based on their legal status. While we are not sure if space warfare will change these distinctions, examining what we use now can be helpful in thinking about the legality of space targets.

The Law of Armed Conflict

In war, targeting is physically limited only by the will of the attacker and his ability to successfully carry out an attack. Despite this, given humanity's horrific experiences with war in the past, we have achieved a remarkable accomplishment in promulgating the Geneva Conventions of 1949 and their subsequent additions.[27] These conventions form the basis of a branch of international law known as the Law of Armed Conflict (LOAC). These agreements, to which states have ostensibly agreed to bind themselves, clearly stipulate what is a valid military target and what is not. They do this not by crafting a long laundry list of items stating what qualifies as military targets and which as civilian, but by establishing broad guidance on how military forces are supposed to conduct themselves. By extension, these agreements *also* direct how *civilians* and noncombatants are supposed to conduct themselves as well. This has major implications for those caught up in wartime conditions; a civilian who in his rage picks up a weapon in the middle of a fight, or who allows one side or another to use his home as a supply depot, has forfeited his protected status. This will likely not change in the space domain.

Legally, LOAC is valid and applicable to interstellar and interplanetary conflicts. The conventions do not set geographical limits, and are clear in their application to all parties of a conflict (state and non-state actors). This means that any fighting which goes on outside the atmosphere is technically subject to these conventions, regardless of how distant the fighting is.

Naturally there are problems with this assumption. While LOAC could one day be the foundation to a larger and more robust agreement on conducting warfare in space, it must first be obeyed by its signatories. Since its promulgation, LOAC has been challenged in nearly every war; the more brutal, the more it has been ignored. Since the end of the Cold War and the subsequent rise of ethnic and religious violence, and ethnic-based independence movements across the globe, groups and tribes fighting for their very existence have found it difficult to comply with LOAC. This is especially true if their rivals are long-hated competitors seen as inhuman opponents.

Applying what we know happens on Earth to a potential conflict between humanity and a non-human competitor produces a bleak outlook. If complying with voluntary restrictions on combat is a problem for human groups, how much more difficult will it be between completely different species who may not even value the other's existence? How could one side expect to trust the other? To hard-liners, it would be like making a treaty with an insect or other pest; the insect could never comprehend the treaty in any case, and even if it could its sinister nature would prevent it from keeping faith. Like the old tale of the scorpion who hitches a ride on the back of a frog to cross a river, the scorpion cannot help but sting, no matter what it promises.

When it comes to space warfare, LOAC provides a good foundation but would require substantial revision to better apply to the space environment. LOAC revisions will be particularly challenged by legal notions of territory which are surely waiting to vex different political groups vying for supremacy in space. Even today, states are having trouble cooperating where sovereignty allegedly ends in high Earth orbit; it is likely territorial claims will be respected solely based on the ability of a state to safeguard said claims. In this early "space wild west" scenario, we can sadly expect LOAC to take a back seat to territorial ambition until a preponderance of force allows LOAC to be imposed on a nascent international space community.

The Simple Sphere: Chasing the Ideal Space Superiority Fighter

Popular science fiction, either through imagination or lack thereof, tends to describe space combat as a mirror of our naval combat here on Earth. Perhaps because the naval model is so alluringly transferrable to space warfare, film after film and story after story relate intrepid space fleets engaging each other from afar via space-faring fighters which not only blast away at each other, but also seek out enemy "carrier" capital ships to bring pain on the opposing fleets. Is this a realistic model? Would naval forces really be carbon copies of our fleets here on Earth? A brief analysis indicates the answer is: probably.

The Primacy of the Spacecraft Carrier

The modern Carrier Battle Group is a direct result of carrier-based airpower and submarines, two war machines perfected and refined for combat as the curtains rose on World War II. As war broke out, many naval theorists and serving naval officers alike were certain the battleship would remain the dominant force upon the seas, and the nation that amassed the most of these dreadnoughts would hold a significant advantage during naval conflict.[28] As it turned out, the British, who certainly favored this idea if they could be called its chief proponent, were among the first to feel the sting of the modern aircraft. On December 10, 1941, the Royal Navy was forced to watch two of its finest warships, the battleship HMS *Prince of Wales* and the battlecruiser HMS *Repulse* be sent to the bottom by Japanese land-based and carrier-based planes—on the same day.[29] Some believers would continue to resist the clear end of the battleship until well into the war, but the undeniable efficacy of air power accelerated the aircraft carrier's place in the limelight, and completely changed naval forces forever.

One of the most significant lessons we have learned about the effectiveness of carriers is that only when carriers are *adequately defended* by concentric defensive layers of *other* specialized naval vessels does their effectiveness become maximized and their survival ensured. Carriers are notably poor at self-defense; despite an onboard air fleet bristling with weapons, carriers today depend completely on escort submarines and destroyers to find, locate, track, and destroy adversary undersea forces. They also depend mightily on other vessels for air defense; while a carrier can certainly hold its own against enemy fighters via its own onboard fighter complement, vessels such as AEGIS cruisers which can get a better total picture of what's going on in the air increase survivability significantly. One ship cannot do everything on its own, even with an onboard population similar to a small city and a large fighter complement at its beck and call. Further, carriers are not exactly small. They require a massive support network, complete with fuel, food, spare parts, and any manner of support a crew of that size needs. Support and resupply methods change drastically depending on whether or not full combat operations are ongoing. During peacetime, smaller naval resupply vessels and contractors on land can be counted on to safely sail to rendezvous with carriers with little interruption. During wartime, resupply efforts must be guarded carefully and executed clandestinely to avoid interdiction; this necessitates different safety procedures which affect tactics chosen by the fleet.

Why does all this matter? In short, the lack of an equivalent "deep space submarine" completely changes the calculus of any future space navy compared to its maritime cousin. The primary reason carrier battle groups look as they do is to defend the carrier, and no small portion of this defense involves undersea threats. With no submarines to counter, the makeup of space fleets, therefore, will more closely resemble a curious mixture of old and new fleets. The carrier will remain the centerpiece and will likely grow to an enormous size, and supporting cruisers, fast destroyers, and even battleships will return to prominence. Perhaps we'll even see the resurrection of the frigate, which has fallen out of favor in the U.S. Navy.[30]

Are carrier-based fighters and bombers still useful in space? The answer is a resounding yes. Carriers will certainly play a role in future space combat. Their onboard spacecraft can be leveraged to increase damage against a potential foe, scout and screen for a fleet in the manner modern aircraft do, and deliver large weapon payloads to enemy fleets without having to close to weapons range with the main vessels, saving the larger and more vulnerable ships from damage or destruction. Further, carrier-based spacecraft will not be subject to firepower limitations which affect earthbound strike craft. Here on Earth, as strike payloads and weapon weights increase, aircraft size and engine power must also increase to allow aircraft to carry their payload safely into the air without falling off their respective carrier decks or careening off the end of runways. Naturally, larger and heavier aircraft require longer runways to reach safe takeoff speed with full loads, and these larger aircraft can also be designed for longer distance flights as their fuel tanks can be proportionately bigger. This physical design limit that still shackles all terrestrial aircraft is well known, and was one of the main issues at stake during the Pacific island-hopping campaign during World War II. Once Japanese islands were conquered, U.S. forces built larger runways which allowed stationing larger and more powerful bombers, like the B-29, within striking distance of the Japanese homeland. Naturally, if these aircraft could have been stationed on U.S. carriers at the time, they would have been; but the deck length necessary to host such bombers, let alone the bomb storage and fuel reserves needed to fly them, was out of reach—and still is—for naval engineers.

The space payoff is this: spaceborne strike craft will not be subject to these limitations. The biggest differences are gravity and atmosphere—the lack thereof. In space, an environment where everything is weightless all the time, strike craft demand no special wing length, no fuel considerations beyond range, and no limits to bomb load except physical space on the craft. Speed will still be a factor of inertia as it always is; more massive craft will take more thrust to initially move, to change direction, and to come to a stop. But once in motion, spaceborne strike craft—and their weapons—will tend to stay in motion without an inertial counter force acting upon them, just like Newton said.

No Weight, No Wings

Another difference between terrestrial and spacecraft carriers is the strike craft themselves. We already know spaceborne fighter craft will not need wings to launch from their carriers; just a good push out the door. With no air to generate lift, spacecraft are in their element as soon as they clear their vessels' hulls. Without wings, designing spaceborne fighter craft could be as simple as a sphere.

The primary obstacle to movement in space is an object's own inertia. This is why spacecraft maneuver using thrusters; they must carry onboard their own capacity to direct their own motion rather than depend on thrust, drag, and weight as in terrestrial aircraft. This means from a pure movement standpoint, a sphere, which has maximum surface area facing actual and potential directions of motion, can be fitted with thrusters and movement devices all over its surface to change directions or halt as quickly as the occupant's safety allows.

When it comes to maneuverability, a sphere in space would be mathematically unmatched. In terms of industrial and military design, however, its function as a weapons platform (which is essentially what a fighter or bomber really is) may not be as exciting. For one, the lack of any flat surface will force the weapons themselves, should they be solid objects like missiles and not directed energy, to be spherical in shape to fit into any exit ports on the vessel. Further, good maneuverability means a baffling and nauseous nightmare for the pilot inside for another main reason: G-Force.

G-Force stressors, familiar to all terrestrial pilots, do not disappear just because one is flying in a zero-gravity environment. G-Force, noted hereafter simply as "g," is the rate of acceleration upon an object as noted by the following equation

$$g = (v_f - v_i)/(\mathrm{t}_f - \mathrm{t}_i)$$

where v_f is final velocity of a moving object, v_i is initial velocity of an object, t_f is final time taken to accelerate to final velocity, and t_i is initial time at which acceleration begins. To avoid confusion, certain equations determining *universal gravitational force* include the component G. This is not to be confused with little "g," which is what we are talking about, but rather is the universal gravitational constant, as determined by Newton.

Even in deep space, *g* affects bodies in motion—and human bodies within those bodies—such as spacecraft. Humans have a threshold for how many *g*s they can endure safely without suffering injury or death. We conduct our daily life on Earth with no issues at 1g.

Critically important is how long it takes to accelerate. A sudden onset of *g* brought about by a sudden control input can be deadly. This phenomenon is called instantaneous acceleration or instantaneous *g*. In space, the current preferred method of flying is to fire rocket or thruster burns for as little time as possible, then "coast" or "fall" towards the destination by using the object's unimpeded inertia. The time spent firing rockets exerts *g* force upon a spacecraft, and this burn must be calculated to make sure the occupants can still function or even survive. Unlike an atmosphere where flying objects are subject to parasite drag, gravity, and weight considerations which all act to impede an object's forward movement, in space only the gravitational pull on surrounding objects affect the flight path and trajectory of objects without thrust (also called "falling objects"). Indeed, deep space navigation models use gravitational nudges and pushes from surrounding planets and objects to help guide craft through the maze of shifting stellar objects over a long flight path and flight time. Getting gravitational influences correct is thus more critical the farther a vessel has to travel.

Gravitational influences can also assist space travel. When one refers to "slingshotting" a vessel around a particular stellar body or into a different azimuth or direction, they are referring to taking advantage of a gravitational influence to change a vessel's course. Without such aid, scientists never would have been able to successfully launch our probes which have traveled to the farthest reaches of the solar system. Voyager 1 would never have escaped our solar system's gravity; and the crew of Apollo 11 would never have made it to the moon and back.

When accelerating in space, careful consideration must be made to how long and how fast a rocket burn fires so a craft's occupants are not totally incapacitated or killed, or the craft itself is not destroyed. This is seldom a problem here on Earth even with our fastest fighter aircraft; the time it takes to accelerate to desired speed at maximum afterburner can subject the pilot inside to high *g*, but this acceleration is limited in two ways: the atmosphere itself (drag), and the engineering tradeoff between engine output, fighter size, and aircraft weight. In other words, terrestrial aircraft engines usually lack the thrust to incapacitate their occupants by acceleration alone; usually a violent maneuver or aggressive dive is required to put aircraft occupants in extreme *g* danger. This latter limitation, engine thrust, is a unique factor of aircraft functioning within a limiting atmosphere: at all times during its flight, an aircraft must fight its own weight, Earth's gravity, and the increasing effect of the molecules found in air as the aircraft flies faster, also called "parasite drag." This means engines must be small and light, which necessarily restricts their thrust as compared to the aircraft's total weight. An aircraft with dangerous engine power could certainly be designed by aerospace engineers, but it would be foolish to do so unless the whole point of the aircraft is to simply go fast.

In space, the atmosphere does not restrict spacecraft in this fashion, and the theoretical size and power of engines mounted to spacecraft is not limited by weight. This makes maneuvering itself a great danger and here again is our friend instantaneous *g*. Basically, the faster a spacecraft goes when it suddenly whips around into another direction or completely alters its vector (composed of a direction and magnitude), the larger the instantaneous *g* value becomes, and the worse the *g* force affects the craft's occupants.

An example helps us understand this phenomenon. When a pilot suddenly jerks back on his control stick and violently alters the direction and speed of the aircraft, instantaneous *g* occurs. In this case, *g* is best expressed in the familiar physics equation

$$\vec{F} = m\vec{a}$$

where F is force, m is mass of the aircraft and occupant, and a is acceleration. In our example, the *g* experienced by the pilot would be the acceleration (a) in the equation. If the resulting force (F) is strong enough, it could rip the wings right off the aircraft. But before that, instantaneous *g* is likely to quickly incapacitate the pilot by causing the blood in the brain to suddenly rush to the lower extremities, as an unfortunate inertial fluid dynamics experiment plays out within the human body.

Why does this happen? Because when *g* affects an object, that object's weight also multiplies linearly with the amount of *g* applied.

$$w = m \text{ x } g$$

Thus, a 200-pound pilot actually weighs 400 pounds at 2*g*, and so on. The body's fluids, primarily blood, also multiply in weight at the exact same rate as any other object on or in the aircraft, and just like the pilot's body. However, the pilot's body doesn't move, and the now-heavy bodily fluids, as fluid dynamics math confirms, are pressed in the direction of the force. In modern aircraft, in a positive *g* maneuver that direction is toward the pilot's feet (assuming they are sitting right-side-up at the controls).[31] The reason the fluids move in the direction they do is that the fluids, like all objects in the aircraft, move parallel in the direction of the force itself—where the force pushes them. This works the exact same way in space. If, for example, a theoretical spherical space superiority fighter was flying straight in one direction at 100 km/s, a sudden change in the opposite direction would exert a force on the pilot similar to a car screeching to a halt. If this is done too aggressively and with too much *g*, the occupant could suffer whiplash or a concussion, or even be flung from their restraints and smashed into the front wall.

Another way to think about *g* is to see the "*a*" in the above equation as simply the result of how fiercely a pilot operates his individual controls. A stronger pull back on the stick means a higher value for "*a*" because the aircraft changes direction faster. As a pilot eases off on the pull, the *g* value would correspondingly decrease as the pilot relaxes his grip. Modern fighter aircraft certainly allow pilots to put themselves in positions where it would require more *g* maneuvers than they could safely endure to recover from a careless or dangerous maneuver. The most immediate consequence of too much *g* is unconsciousness as the blood is pushed toward the feet and out of the brain, leading often to an aircraft which continues its dangerous course or goes out of control altogether. The biggest challenge after too much *g*, then, is to survive and regain consciousness in time to recover the aircraft into stable flight. In this same example in space, the pilot would eventually—hopefully—wake up to find his craft careening in an unknown direction, likely quite far from its original destination.

In deep space, not only is instantaneous *g* a concern, but so is simple acceleration of a vessel of any kind, with no turns and no fancy maneuvering. This life support problem is best understood with a large-scale example. A mythical spacecraft carrier is transiting from the moon to its on-station patrol around Mars, a distance of about 55 million kilometers at its closest orbit.[32] The vessel has orders to arrive in six months, which requires an average speed of 1300 kilometers per second (km/s). To reach this speed, the spacecraft will have to accelerate to it. Next, we must find the mass of our carrier. Since $m = weight \times gravity$, we're in luck: the mass of the carrier is equal to its weight, since gravitational force in this case is negligible in deep space.[33] Thus we can use the arbitrary value of 200,000 metric tons—our fictional carrier's displacement—for our example. Now that we have the right data, we can start finding the right *g*. If we first try an instantaneous acceleration to 1300 km/s, where we reach cruise speed in one second, we can solve for *g*:

$$g = (v_f - v_i)/(t_f - t_i)$$

Once we plug in the data, we get

$$g = (1{,}300{,}000\text{m/s} - 0)/(1 - 0\text{s})$$

which yields a *g* of 1,300,000 m/s.[2] Such a force exerted on the occupants of the carrier would not end well. The trick, then, is to increase the *total* time; that is, the time it takes to accelerate to a desired speed. This means the best way to solve this problem is to select your desired—and safest—*g* first before hitting the afterburners. If we choose the most comfortable *g*, 1G, this would mean our carrier would essentially accelerate without our occupants even knowing it; the crew would feel as though they are standing still on Earth at 1G. Thus

$$1 = (1{,}300{,}000 - 0)/(t_f - t_i)$$

In this case, the math is made easy. For an acceleration at 1G from a standstill, simply accelerate at 1G for the same value in seconds as your desired speed. This means 1,300,000 seconds, or approximately 361 hours, which is about 15 days at a constant 1G acceleration. If the crew is willing to accelerate at higher rates to get to cruise velocity faster and potentially conserve a bit more fuel, the equation scales linearly. At a 2*g* acceleration, only 180 hours of burn are necessary, which takes about half as long—about a week.

These calculations reveal several problems. First, for the week or two you are *not* travelling at 1300 km/s, you'll need to make up the difference by increasing speed slightly faster than the target speed to arrive at the destination on time. Second, slower *g* accelerations require a longer fuel burn at lower rates to get to cruise velocity, which could be a fuel concern for smaller vessels. Third, asking a crew to sustain *g* loads higher than 1G for days on end will carry with it significant health concerns, and is probably not medically feasible, nor is it smart leadership. Finally, and perhaps most importantly, a spacecraft engine capable of accelerating consistently at 1G per second for days on end first needs to be invented; as of this writing, none exist.

The tactical and strategic implications of space travel when planning for *g*, and not for a best speed, are significant. First, since any vessel will be mathematically weightless, this means if two or more vessels embark on an objective at the same time from the same place, and accelerate at the same *g rate per second* to the same speed, they will arrive mathematically at the exact same time. This will be true regardless of vessel size, shape, and complement. Travel like this will never be perfect as mistakes will likely occur with engine thrust, miscalculations or overlooking nefarious gravitational effects of nearby stellar bodies will interfere with travel routes, and other issues are sure to pop up. Nevertheless, speed, design, and *g*'s implications on strategic deployment and tactical concentration of forces are clear.

Tactical and Operational Considerations of System Assault

Planetary systems closely resemble medieval castles or fortresses in their ultimate strategic function. A force desiring control of a particular sector of space or series of systems will be forced to besiege and take planetary systems which are in the way of its territorial ambitions. Why? Because, just like forts of old, a system bypassed by a military force remains a threat if it happens to be in the general vicinity of military operations. In space, the "general vicinity" is vast indeed. The differences between castles and star systems in this regard are only by degrees of distance and speed; at any point, enemy forces in a bypassed system can sally forth and harass an enemy's supply lines, suddenly extended after bypassing the garrisoned system. Worse for the attacker, unlike castles of old a stellar system allows relatively untrammeled rest or refit operations, and in the case of planets has the potential to be self-sufficient and therefore impervious to traditional starvation tactics. A system held by an enemy grants them opportunities to harass or plunder forces travelling near that system, and thereby extend and supplement its own supplies at the expense of the enemy. Worst of all, unconquered systems could contain unknown numbers of concealed enemy forces.

Once a decision has been made to attack a system, operational planning should immediately concern itself with the approach to the system. Of all the components of the approach, the most important will be how to set up the assault on the enemy's possessions within that system—the direction, relative azimuth, and flight paths required to secure objectives in the system.

Basic navigation in deep space is tricky enough without worrying about how to approach something relatively undetected. While systems seem simple from the outside, as it turns out a system is actually a briar patch of electromagnetic flora that can offer interesting advantages to an assault force.

Multiple Suns, Multiple Hiding Places: Tactical Concealment in a Star System

In December 2013, the Cassini-Huygens space probe, a joint venture between NASA, The European Space Agency, and the Italian Space Agency sent a probe to Saturn to study its unique rings and surrounding moons. What it found, according to NASA, changed the way we think about the planet.[34]

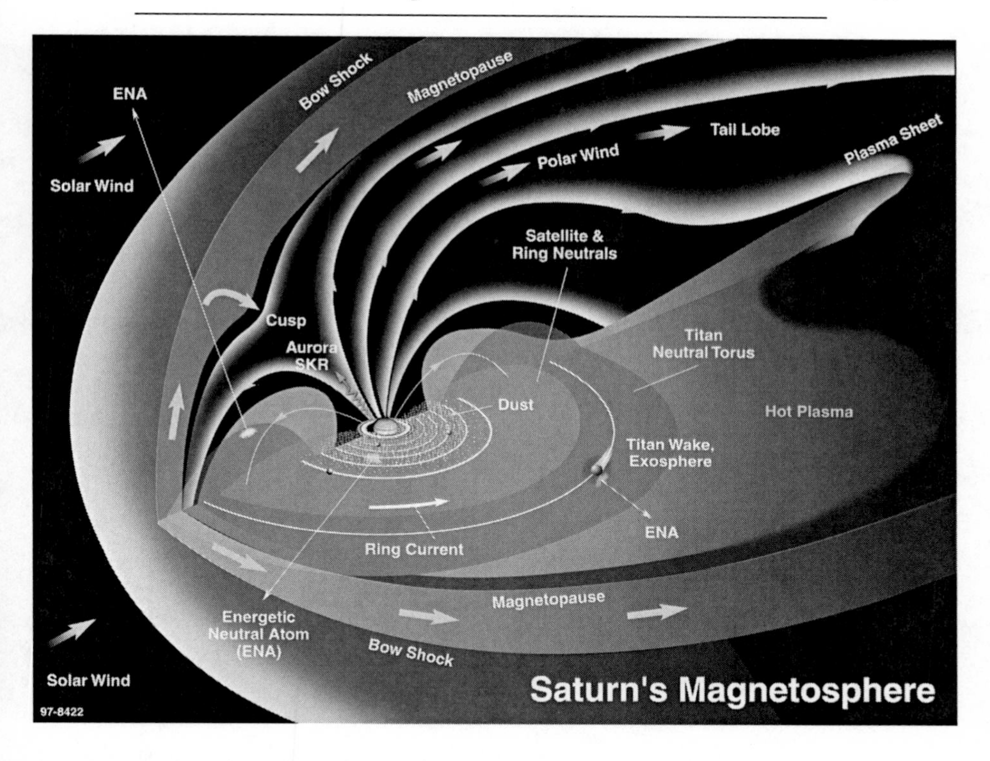

Figure 8: Saturn and its surrounding EM features[35] (National Aeronautic and Space Administration (NASA), website: https://solarsystem.nasa.gov/missions/cassini/science/magnetosphere).

What was once thought to be a relatively benign gas giant was discovered to have a robust magnetosphere which influences all stellar bodies near the planet. According to NASA, this magnetosphere emits its own radio waves, and also blocks radio waves from outside the sphere, which means Saturn's magnetosphere could not be studied by scientists on Earth until the Cassini probe physically approached it and breached its surroundings. Saturn's atmosphere-free sixth moon, Enceladus, feeds this magnetosphere by losing its surface water ice to Saturn's gravitational pull. This ice finds its way into Saturn's rings, where it is converted into plasma, the primary source of the magnetosphere.[36]

What does this mean to an interstellar operational planner? The existence of Saturn's magnetosphere means there are surely more like it around other planets; and one of the magnetosphere's primary attributes is that it essentially acts as an electromagnetic oasis which allows those inside it to communicate and signal without outside interference. Essentially, background radiation, radio waves, and the EM spectrum which one would normally read *outside* Saturn's magnetosphere would read completely different within the magnetosphere. Without the proper calibrations to analyze radio waves and emissions accounting for this internal-external discrepancy, any force hidden within this magnetosphere would be for all intents and purposes electromagnetically shielded and therefore completely hidden from external detection. This makes places like this exceptionally good for fleet rally points, astrosynchronous stations, and muster locations to prepare for a planetary assault.

Expanding upon this, an assault force, properly situated and distant from the target planet, could potentially "oasis hop" from planet to planet until they either run out of oases or are within striking or detection distance of their target. Much depends on the ability of the enemy to detect an approaching force, reaction time, and their fleet's capabilities. In fact, for a sufficiently inferior enemy, "oasis hopping" may be futile and wasteful if the enemy can't even detect the assault fleet in the first place. This fact simply underscores the need for proper intelligence preparation prior to launching the assault fleet.

"Oasis hopping" can be further enhanced by approaching from a direction with a large body of stellar objects, like an asteroid belt, or from a stellar phenomenon which naturally produces confusing readings, like a nebula or black hole. One of the best possible system approaches could look something like this.

A fleet admiral begins his approach to the enemy system beyond the enemy's known surveillance capabilities. Before hitting this surveillance boundary, he maneuvers his force through a nebula which extends beyond and into the enemy's maximum detection radius. At this point, the admiral directs his forces towards the nearest gaseous outer planet of the enemy's system, hiding in the electromagnetic oasis of the closest planet to the nebula. The time the fleet is exposed within the enemy's detection radius is minimal, and unless they were looking for them, he knows he would not be found.

Next, the spacious expanse provided by the electromagnetic oasis allows the admiral to deploy an astrosynchronous base, allowing his vessels to resupply, muster, and refit as necessary. The station remains as a forward base and will hold extraneous equipment to slim down the assault force, acting as a spacegoing equivalent of the soldier dropping his camp gear and pack and bringing only his shield and sword to battle. Once the assault begins, the base will remain the field headquarters until a suitable one can be established in orbit around the target.

Once sufficiently prepared, the fleet admiral leads his forces to the next-closest electromagnetic oasis provided by the nearest planet generating one. As the force approaches an asteroid belt, assuming one is available, separating the outer solar system from the inner, there is only one tactic left: concealment amongst the rocks. As the vessels begin picking their way through the asteroid belt, enemy sensors can easily confuse the strike force with the rocks themselves, offering the assault force this one last chance at concealment. Upon breaking out of the asteroid belt, the race begins: launch the attack before the enemy detects the assault force and musters its defenses in an organized fashion. If successful, this sudden appearance of a massive assault force will likely engender a helpful amount of enemy panic and tactical surprise.

Electromagnetic Oases: The Problem of Intra-System Surveillance

Conversely, such electromagnetic oases can also stymie intelligence-gathering efforts. If an assault force can hide in it, so can the enemy. This makes them excellent spots for system defense: astrosynchronous fighter bases, listening posts, automated weapon platforms, and any assorted booby traps which could be cooked up by a determined foe.

Some kind of electromagnetic disturbance, like Saturn's, will naturally occur with every planetary object in a system, not just large gas giants like Saturn. A key to planning an assault is to first understand these disturbances, their potential tactical advantages, and their potential for enemy deception. An assault force should expect conflicting data, confusing readings, and contradictory electromagnetic intelligence reports to be par for the course. From what we learned from Cassini a mere five years ago, it is a safe bet that electromagnetic phenomena exist which we still do not understand, and which will both act as intelligence havens and disruptors to those seeking to keep a low-profile during system maneuver and tactical assault.

The Ephemeral Advantages of Automated Weaponry

The above analysis makes a seductive case for automated forces, vessels, and weaponry of any kind. The *g* limits discussed naturally incline one to think robots could do better over squishy humans. Further, the long distances and supply requirements which come with space dominance tempt one to also think relying on robots could solve most problems, from energy consumption to food to vessel design concerns. Complicated maneuvers to ferry an assault force from electromagnetic oases through an asteroid belt and beyond could easily benefit from a machine's cold precision and discipline, not to mention radio silence and emissions control.

Despite this, the future space admiral should take caution in this approach. While it may seem convenient, one should never underestimate the power and confusion of the hunt as the forces approach the enemy. Clausewitz, for his part, immortalized this fact with his terms the fog and friction of war.[37] A totally automated force removes the human element—the ability we possess to think on our feet, adapt, incorporate or discard new and rapidly developing data, to feel our way through the fog and slough off the friction. A machine can only do as it is told, and can only improvise within its programmed parameters. No doubt machines will be necessary to assault a future spacefaring enemy, but placing them in charge would be a mistake.

When close encounters are imminent, the human race can always be counted on to do what it does best: close with terrifying speed and determination at a foe, using all its guile and force to unseat an adversary from his perch with vigor and destruction. In this chapter, that has meant space dominance—the pursuit of a preponderance of military power in a particular region of space which allows us to successfully utilize military force at a time, degree, and place of our choosing. To achieve this dominance, a military force must consider spacecraft design, fleet composition, travel and acceleration concerns, operational approaches, and the characteristics of space and its residential anomalies.

This approach to dominance, while messy, is reliable. It is also, sadly, familiar. It has been part of the human condition for several thousand years, and will continue to be just as applicable in the future. For all we know, the human race could be the biggest, most violent, most terrifying species that exists in our galaxy; after all, we are both willing and capable of ruthlessly obliterating entire populations of enemies on our own world, consisting solely of our own species, while simultaneously damning the environmental and social consequences. Other species, upon learning about us, may come to view us as bloodthirsty savages whose unquenchable thirst for destruction will lead them to give us a wide berth. If they fail to analyze us correctly, they may one day hazard to find an assault fleet in orbit over their home world, with a strange species of primates ready to make no distinction between combatant and innocent. That will be a dark day indeed for that enemy.

5

Planetary Invasion

> In a mixed ship [men and women] the last thing a trooper hears before a drop (maybe the last word he ever hears) is a woman's voice, wishing him luck. If you don't think this is important you've probably resigned from the human race.
>
> —Robert Heinlein, *Starship Troopers*

The amphibious assault reached its zenith in the 20th century. Following false starts and failures in World War I, the next war saw ambitious amphibious assaults of unprecedented complexity, scope, and length. Germany, Japan, and the Allies all used amphibious assaults with impressive results. Skill in conducting amphibious assaults gradually grew and culminated in the Allied "island hopping" campaigns from 1942 to 1945, where U.S. forces reached a professional proficiency in planning and conducting amphibious assaults never seen before in human military history.[1] During each operation, with the help of prolific home front war materiel production, military planners ensured the assault forces had a preponderance of weapons, supplies, and above all, assault forces which outnumbered the enemy garrisons. Even with stubborn Japanese military and civil resistance, during this campaign not a single defensive operation succeeded in throwing an enemy back into the waves.

It is clear it takes more than determination to hold an island when it is swarmed with fresh, well-supplied veteran assault forces. The lessons learned from these assaults produced an impressive combat record: of all the major amphibious assaults made by *any* belligerent in World War II which targeted islands, not a single assault failed.[2]

This fact has major consequences for planetary invasion. A planetary invasion, after all, is just like an amphibious assault on an island; except the island in this case is a floating planet amidst an ocean of stars. This lends credence to the idea that if a planetary invasion campaign is sufficiently supplied, armed, reinforced, and pressed home with determination, the odds of its success are much higher than the defenders' chances of winning.

Now that we have discussed the effects of ideology, logistical requirements, and achieved space dominance, we can begin tackling an operation of almost unbelievable magnitude: planetary invasion.

No doubt most people's concept of planetary invasion comes from science fiction novels and movies. But once the glittering trappings of an alien horror film have subsided, the realities facing such an undertaking are monumental. Marshaling and equipping the forces will actually be the easy part; the most challenging aspects to planetary invasion include determining the right strategic objectives, operational approaches, and tactical forms of attack, all complicated by the vast amounts of time required to conduct such an operation, including travel time to the target. The task is at once as baffling as it is intimidating. In fact, many would properly ask *why* an invasion should be undertaken in the first place, when an invading force could simply pulverize a target planet with standoff weapons from a distance.

As history has shown us over and over, total destruction of an enemy via violent military attack is rarely possible, nor does it guarantee achieving one's political objectives. This book addresses planetary invasion as both an interesting military challenge and eventual military possibility. By investigating deep space warfare from the perspective of the most tactically complicated kind of military operation imaginable, by extension less complicated military operations become clearer.

Taking a planet will be no mean feat. Planets come in many different forms, and the purpose and end result of a successful invasion needs to be considered before the assault can be planned. To aid planning, there are three general categories which describe potential planetary invasion targets:

- A planet already inhabited and habitable only by the enemy. Habitable planets are in many ways self-contained units, and planets capable of supporting enemy life are likely also capable of sustaining a garrison indefinitely. Aside from purely military articles like ammunition and equipment, a planet amenable to the enemy's preferred living conditions but hostile or toxic to ours poses the most difficult threat our forces could face. This category only applies in the case of war with a sentient non-human species. During an assault, every square inch of such a target will be contested by both enemy forces and the environment itself, and environmental protection requirements for our forces will complicate combat operations. If the campaign progresses favorably, the local population will become more and more desperate to survive, and desperation will power their resistance. Complete extermination of the planetary population may be required to fully pacify the planet; this is a major ethical consideration which must be understood before committing to an assault.

- A planet or planetary body uninhabitable by any combatant. In this case, the target planet could be a large variety of possible environments, all of which are inhospitable to life as we know it, and the lives of any non-human combatant. The reason for taking such a planet would be to deprive the enemy of a strategic vantage point, any installations or satellites in orbit around it, or an object of cultural, religious, or intangible significance to the enemy.
- A planet inhabitable by both species. This condition is essentially no different than the first case, except that it offers our forces the benefit of a friendlier environment and allows human assault forces to save time and burden themselves less by forgoing the need to bring along cumbersome environmental protection gear. Proper examination of the atmosphere and resident flora, fauna, and microbial life will be essential prior to allowing human forces to operate without environmental protection.

When planets or planetary bodies need to be taken, one of the above three types of planetary assault conditions will dictate how the campaign should be planned. Though the task may seem intimidating, depriving an adversary of a settlement, an industrial center, or even his homeworld will impose crushing repercussions on an enemy war effort. In a limited war, seizing enemy planets is an excellent way to bring them to the bargaining table.

Using humanity's experience with amphibious assaults as a model, we can discern several clear phases in any planetary assault:

- Stage I: Blockade and strangulation. Controlling complete access onto and off the planet is essential to beginning military operations on and above the world. While it may be difficult to picture, a three-dimensional blockade above the planet would be necessary to proceed with further siege operations. This three-dimensional blockade means vessels stationed above key spaceports and on the dark side, light side, and polar (if applicable) aspects of the target planet. Communication satellites should be appropriated or destroyed; traffic to and from the planet highly regulated. This phase also counters the argument a planet can be reduced from a distance with standoff weapons only; if enemy vessel freedom of movement around and near the planet is not subdued, be it military or civilian supply vessels, standoff weapons may not be enough to do the job on their own as the target planet is repaired and freely resupplied.
- Stage II: Planetary Siege and Orbital Bombardment. The planetary invasion-equivalent of "softening up the target," bombardment from orbit has several excellent advantages for the assaulting force. First, the vantage point from space gives gunners excellent targeting solutions for stationary and strategic targets below, such as government centers, spaceports, military

bases, fielded forces, and industrial zones. Second, orbital bombardment does not depend on ballistic trajectories which hamper traditional artillery strikes. This means less shots go wasted on incorrect fire calculations, wind resistance, weather, and so on. Third, orbital bombardment is largely unaffected by terrestrial weather, time, or day or night conditions. From their perch above a planet, orbiting gunships can obliterate targets as they find them, and can continue to aid friendly forces once the assault begins. Fourth, orbital mastery will squelch planetary land, air, and maritime surface resistance by providing a higher ground from which to attack the planet's garrison and defending forces. This provides good cover for any operations the besieging force would like to conduct on the surface before the invasion begins, like prisoner snatching operations, special forces insertions, and surgical strikes which cannot be done from orbit.

• Stage III: Biological and Chemical Warfare. Once an orbital bombardment has yielded military success and the way is partially paved for invasion, any assault force needs to make a big ethical decision. Even if the species under siege is not our own, there will likely be ways to exterminate the planet's population via biological and chemical weapons. If the enemy population happens to be non-human, there may even be chemical or biological means to do so with weapons which do not adversely affect humans. Assuming this non-human enemy is also a carbon-based life form (as they most likely will be), the chemical building blocks—elements of the periodic table—which were used during their evolution will be driven by their preferred environment. The environment and these building blocks will in turn dictate which chemical elements are toxic to the enemy, and since we know what will and will not kill us, appropriate biological and chemical weapons can be fashioned to use on the enemy. Naturally, this is a major ethical decision, and many factors will need to be considered before employing these hideous weapons against an adversary. Suffice to say, this stage can be skipped if found ethically untenable. Nevertheless, it is easy to see a commander in desperate straits, under pressure from the government and presiding over a poorly-run invasion, turn to these horrific weapons as a way of accelerating victory.

• Stage IV: Orbital Insertion and Spacedrop. Once orbital bombardment has softened the planet up as much as it can, when all possible outcomes have been analyzed to the best of the commander's ability, and all conceivable preparations completed, or if the threat of an enemy relief force becomes a pronounced possibility, it is time to assault the planet. This means putting ground forces on the surface and eliminating or subduing all enemy resistance. A three-dimensional blockade around the planet along with space dominance will allow assault forces to simultaneously hit the planet's surface at roughly the same time across the entire surface, which is a great tactical advantage.

Forces Required

Planetary invasion is, without a doubt, a logistician's nightmare. How exactly does one plan for the huge number of forces needed to trample upon an alien world and beat it into submission? How does one plan for a several-million-mile supply line? Upon arrival, how reliable can the life support systems be for the forces launching themselves at an alien world likely unfriendly to human life? How many casualties should one plan for, and how much attrition should be expected to occur during the millions-of-miles cruise? And just exactly how many munitions are required to reduce a single continent, let alone several?

These questions are enough for any logistician to throw their hands into the air and quit. Here we can only describe the way a campaign could unfold, then hope future logisticians can work backwards to complete their unbelievably complicated job. Logistical supplies will largely be driven by the type of attacking force which is selected for the assault. As we stated earlier, our options are a total human force, partially robotic, or fully automated.

Once this is done, logistical planning can begin. It should be no surprise that the vessels required to haul what would be needed for an assault the millions of miles necessary to get there will likely number in the hundreds. Foraging is out of the question; there isn't much to eat between Earth and the target system. Planning for a planetary invasion is certainly a challenge, but in some ways is only different in scale compared with other historical amphibious assaults and invasions.

The Astrosynchronous Base

One of the best ways to assist an invasion force is to establish a base within striking distance of the target, preferably concealed from the target planet. In the case of a planetary invasion, there are several advantages the attacker has when establishing a forward operating base.

While one's first inclination may be to establish this base on something solid which happens to be nearby like an asteroid or planetoid, that may not be the best tactical choice. Even if a suitable planetary body is not found near the target which can be concealed, like a small moon around the target planet, space allows us the flexibility to construct a base of our own nearly anywhere.

An astrosynchronous station which is strategically placed far enough from the target planet to avoid surveillance but is still a quick hop away is certainly in the realm of the possible, and may even be required for our first attempt at planetary invasion.

Second, no matter where it is located, such a base would allow for total-force mustering (i.e. concentrating the entire invasion force in one location), act as a staging area, and also as a safe haven where repairs and refits can be conducted prior to major operations. Third, a base like this acts as an area for withdrawal and a rear supply station, which can house the wounded and reload armaments as needed. Finally, when the major combat dies down the astrosynchronous base can be moved or converted to a geosynchronous or geostationary base above the target world, acting as a floating military governor's palace which can monitor the conquest process.

Endgame Objectives

Before undertaking any great task, it is always best to begin with the end in mind. The Prussian military theorist Carl von Clausewitz essentially divides war into two major categories which still exist today: total war, which is the complete capitulation of the enemy, and limited war, or war for the purposes of obtaining a particular objective or objectives.[3] Deep space warfare will not change the nature of war; however, it is worth briefly discussing how its character could be shaped by these two categories.

Total war offers little in the way of discussion. It is, as French military leaders once called it, *guerre l'a outrance*—a war with no limits, where the attacker holds nothing back in his desire to stomp the enemy out of existence. It also requires total mobilization of a society for warfare, and subjects an entire nation's produce and energy to making war. In total war, everything the enemy is or owns can be a target; combat tends toward dehumanization (or de-sentientization in case of a non-human sentient enemy) and state-sponsored hatred of the enemy. Horrific decisions resulting in death and destruction on a wide scale are authorized and sought by military and civilian leaders. In total war, an entire group's energy is directed towards the eradication of another. Such activity is never pretty.

The most likely type of deep space war, and also the most realistic to execute, is limited war. While total war strives to completely eliminate a chess piece from the board, limited war attempts to place said chess piece in a compromised or controlled situation which allows a particular political group to accomplish what we set out to achieve. Limited war seeks limited gains, but those gains have been fully vetted as political objectives supported by a feasible military strategy. Limited war does not seek total destruction of an enemy, but could seek total capitulation of a particular group.

Given this, we can divide endgame objectives into three major levels which correspond with the levels of war: strategic, operational, and tactical. These objectives build upon each other, ultimately requiring a chosen military strategy to achieve a political goal.

Strategic Objectives

These objectives are the big-picture, national level reasons for going to war. Strategic objectives represent a clear statement about what they are willing to commit precious military resources to obtain. To allay confusion, strategy discussed is this chapter is military strategy, which is *not* the same as grand strategy. When the military is chosen as the appropriate instrument of power for achieving a political objective by a government, only then should military strategy address just how that objective should be achieved by military power.

A strategist needs to take care not to misidentify lower-level operational and tactical objectives as strategic ones, with the understanding that tactical and operational actions can have strategic effect. When we say "strategic," what we are discussing is effect, and not level of warfare. That is the key difference: all levels of warfare aid the ultimate political objective, and when military power is chosen to achieve this objective military strategy is simply the "roadmap" to get there. Conversely, when a political objective is poorly defined or resists meaningful measurement, practitioners at the tactical and operational levels tend to strike out on their own and take actions which also tend to fail to support the ultimate strategy. Some examples of military strategic objectives include the following.

- **Eliminate enemy ability to project force onto Earth assets or into Earth-held territory.**

This objective is large in scope, unspecific, and targets enemy military capability writ-large, making it an excellent strategic objective. How this objective could be accomplished with military force is best left to military leaders who know better about their force composition and capabilities. Regardless of tool chosen, though, removing the military threat to Earth is the goal, which supports a political objective seeking better security.

- **Ensure enemy planetary strongholds and sources of military power cannot coordinate.**

This could mean many things, but clearly leads to a more secure humanity if the adversary is particularly threatening. It could mean physically blockading individual planets to prevent their contact, or interdicting enemy supply convoys, or jamming or cutting off enemy communication. Any of these and others are negotiable as long as the objective is met.

- **Coerce the enemy's political leadership into surrendering.**

Indicating the enemy's political leadership is a persistent and primary target reinforces the idea that this campaign will focus on hastening an enemy's surrender, and will do so by striking or coercing the entities capable of negotiation. How and when is left up to other planners; designating them a matter of national interest will suffice for this level of warfare.

Operational Objectives

To support chosen military strategic objectives, operational objectives are set to designate "large muscle movements" which can be obtained exclusively through military action. In other words, operational objectives represent the "big arrows" on campaign maps which designate which unit goes where, and to do what. This is perhaps the most difficult level of warfare to navigate, since operational planners need to understand strategic objectives and also know what tactical resources are available to accomplish then. In many ways, operational planners are those who must responsibly commit resources to a bigger plan, which gives them a large amount of influence over a campaign. Above all, operational plans consist of a *campaign*: a limited set of engagements designed to achieve military goals within limits prescribed by strategic planners and commanders. Some examples of operational objectives are as follows.

- **Disrupt enemy logistic resupply capabilities with the 3rd Fleet.**

In this example, 3rd Fleet was chosen to disrupt enemy logistics likely due to some characteristic or special capability attached to its forces. For instance, perhaps the 3rd Fleet commander best understands the enemy's logistical strategy and would therefore be best suited to counter it. 3rd Fleet could possess a piece of technology or unit which specializes in logistical destruction. In any case, the operational planner must designate in general terms who faces down who, and more or less where they should go at the opening of the campaign. Note there are no other specific details, meaning the fleet commanders responsible for this objective are allowed to get creative as to how they achieve it.

- **Execute a blockade of enemy planets and planetoid strongholds.**

Again, this objective specializes a particular action without getting into the gory details of which unit is on station at what time, how many ships are needed, and so on. A blockade would also pressure the enemy to the negotiating table if the blockaded planet's population or purpose depended on off-world support, which would be a strategic effect of an operational objective.

- **Attack and destroy enemy shipyards orbiting planet Z and Q.**

The direction is clear and the objectives are general, making this a good operational objective. Operationally, the shipyards at Z and Q are clearly more important targets than shipyards elsewhere, hence their inclusion by name. There is clear support for a larger strategic goal, which is likely eliminating the enemy's ability to project military power onto Earth assets and into Earth territory. The details on how the shipyards are destroyed and with what individual vessels and forces are left up to the commanders, as they should be.

Tactical Objectives

The lowest level of warfare (in a hierarchy, not in importance), tactical objectives are also the most consequential ones in warfare, where failure is heavily punished and success often hangs on the decisions of an on-scene commander during active fighting. This is the nitty-gritty level, designating individual units to strike individual targets or take down individual enemy objectives.

- **Seize high orbit, bombard, and assault planet Z with the 5th Invasion Fleet.**

The objective specifies which unit will strike which target, and what action they are intending to do. This objective clearly supports operational and strategic objectives, but still grants sufficient authority to on-scene commanders to achieve the objective without too many constricting specifics.

- **Seek and destroy enemy 8th Fleet with 5th Fleet space superiority forces.**

Again, this objective specifies a particular unit to destroy another unit to support a higher objective. The friendly 5th Fleet in this case could be part of a larger sweep through the area encompassing several fleets, or could be clearing out the enemy 8th Fleet for future tactical gains.

Once objectives have been determined and commander's intent made known, the fight can begin. A closer look at the four stages of planetary assault reveal these objectives in action.

Stage I: Blockade

Any good planetary assault on an inhabited world must begin with a blockade. There are some who may believe this may not be necessary; after all, if we are using an amphibious assault on an island as an approximate analogue for planetary assault, plenty of islands have been taken without blockades in terrestrial military history. Planets, though, are different; especially those which are capable of sustaining their populations, much like Earth would be if it were under planetary siege.

A spaceborne blockade is subject to the same problems a terrestrial maritime blockade is. For one, there's no such thing as a perfect blockade, even with advanced sensor technology and swift vessels. For another, it is quite difficult to cover the entire surface of a target planet with the watchful eyes of a friendly fleet, even with several fleets in orbit. The intricate scheduling and interweaving pieces and parts to a blockade necessarily creates inefficiencies and mistakes, allowing the enemy to take advantage of errors to sustain the planetary garrison. Thus, the best blockade targets are planets or planetoids which host facilities and populations particularly susceptible to economic strangulation and which depend on outside supply for daily operations.[4] This is especially true if the enemy planet has reached a level of sophistication which essentially obliges it to depend on external settlements or space-based industrial sources for its livelihood. Better yet, blockade is very effect if the target is a military facility on an inhospitable planet or planetoid which requires resupply to guarantee the garrison's survival.

Once an effective blockade is set, work to reduce the garrison and defenses can begin.

Stage II: Planetary Siege and Orbital Bombardment

Blockade by itself cannot not win the planet unless the garrison has been so careless as to completely forget to store supplies or if the environment itself compels a surrender. Without such luck, a siege is unavoidable. Humanity has been laying sieges to cities since before recorded history, and from this experience we already know sieges only end in four ways: the attacker gives up, the city is starved into submission, the city is seized by treachery, or a city is successfully stormed through an assault. When the target is an entire planet, there is no evidence which leads us to believe these four outcomes will be any different than city sieges here on Earth.

Thus, every siege requires the attacker to review the possibilities of these four outcomes. The first thing a besieger must consider is if starvation is even possible. New settlements could be sufficiently unstable to still depend on supplies from their homeworld, but it is likely the urgency of a siege could compel populations to creatively find ways to stretch their supplies or develop methods of self-sufficiency. The same is true for large and sophisticated homeworlds; faced with certain destruction, the enemy should be expected to revamp what resources they have left to support a long holdout. While this would likely entail completely retooling their economy, this is not an unreasonable decision when faced with certain demise and imminent destruction by a force beleaguering the entire planet.

If starvation is possible, it should be used first. Not only is it the least messy course of action, but it also stands the best chance of forcing a bloodless surrender by the garrison. Small planetoids with hostile environments, greatly outnumbered military outposts, freestanding astrosynchronous installations and space stations are all good targets for starvation while under siege. In all these cases, the clock is against the defenders, as they stand to be completely overpowered by the attacker's force, lack inadequate defenses, or stand to run out of energy or other life-support materials without resupply.

Military history shows that defenders in a siege are only successful if they execute an active defense, and if they receive timely relief. From the defender's perspective, only one of these things is within their control, that of an active defense. Besiegers should therefore expect spirited resistance by any defender at any location under siege, and must remain at full combat readiness throughout the siege.

This brings us to orbital bombardment. Target selection during a planetary siege will be a point of contention between military commanders, and it is not immediately clear what the best strategy for reducing a planet could be. USAF Colonel John Warden elucidated a concentric ring model for assaulting an enemy state with airpower, advocating its use during the 1991 Gulf War between the United States and Iraq.[5] This model is clearly applicable to an entire planet, and the differences between a planetary bombardment and the Warden-inspired "Instant Thunder" operation in 1991 are only different by degrees, and not in concept. Regardless of the overall strategy chosen to bombard a planet, it is clear targets will encompass planetary defense assets, planetwide communication networks, energy production facilities, military bases, launch facilities, fielded forces, political targets, and targets which allow citizenry to mount organized defenses. If a siege drags on long enough, food production and population sustainment must be destroyed to hasten the planet's surrender.

This last target is particularly challenging on a planet-wide scale. While it may seem a wasteful and insurmountable task to seek and destroy every fertile field from orbit, or lay waste to every military and government installation on the planet's surface, these may be realistic albeit tedious options during a planetary siege.

Lastly, subterfuge is an unlikely solution to planetary sieges. While treachery is an option for terrestrial sieges, it is difficult to see how an entire planet would succumb to a treacherous act and open its proverbial gates. With smaller targets, like military installations, mining outposts, free-floating stations and listening posts, the size of the garrison could be compelled to surrender through treachery if the besieging force is of sufficient size and skill to storm the facility once the treachery occurs. Once we get to planet-sized targets, though, the sheer size and dimensions of a planet make this nearly impossible. On a planet, there will always be those who are unaware of the treacherous goings-on, who remain oblivious but committed to defending their territory to the death, even if ordered to lay down their arms. Speed is also a problem; terrestrial sieges benefit from treachery because once the gates are open, it only takes a single day or night to conquer the place. An entire planet will require much longer. Finally, if the besiegers destroyed enemy communications networks in the first stages of the siege as they should, then communicating the fact that a planet has surrendered to an entire population would be impossible. Indeed, fighting should be expected to last long after landings take place as the population slowly comes around to the fact that they have lost.

Stage III: Biological and Chemical Warfare

World War I saw the introduction of chemical weapons in human warfare. Mustard gas, a crude but effective toxin, was used as a human insecticide by both Allied and Central powers with almost cavalier attitudes during the global conflagration. The horror and injuries caused by these weapons caused most of the world to agree to ban them completely, and for the most part chemical weapons have disappeared from modern battlefields and state stockpiles.[6] When they have been used in the post–World War II era, they have been met with global condemnation, retaliatory strikes, and have almost exclusively been used by brutal and cruel totalitarian regimes desperate to maintain their power. Terrorist groups seeking these weapons continue to be a major global security concern.

Our experience with chemical weapons seems at first blush to suggest our use of them in future planetary warfare should be a resounding "no." However, one needs to soberly examine the reasons why they would be employed and the pressures the human race may be under during an interstellar war, especially if the enemy is a non-human sentient race bent on our destruction. To be sure, it is safe to say these weapons will not be employed against human adversaries without similar human political repercussions as may happen today. But when faced with a non-human sentient enemy, considerations become more complicated. If a fight with an extraterrestrial enemy bent on our annihilation as a species begins to go badly, unleashing chemical weapons is not a forgone conclusion. If the very survival of our species is at stake, would it not be foolish to keep a particularly effective weapon holstered as we are conquered and exterminated? Moreover, as previously noted above there may exist chemical or biological weapons which affect a particular enemy and not humans, making the use of such weapons that much more intriguing.

It is not hard to imagine an admiral or general employing planet-wide chemical or biological weapons which are designed to be short-lived, but with devastating effect on the enemy's military and civilian population, leaving gaping holes in their defenses to be easily exploited by human assault forces. As with any weapon, the unknown effects sure to be caused by the use of chemical or biological weapons on a planetary scale should be carefully considered by any force willing to use them; as should any resulting political consequences. If humanity is pursuing a limited war with the enemy, using chemical or biological weapons would only complicate any peace negotiations or enemy submission, not hasten them. This is because the use of these weapons would likely be perceived as a heinous crime by the enemy, and drive them more towards perceiving humanity is attempting to conduct total war rather than seek limited concessions. Then again, it may not.

If there are any political or military doubts, or if the war is not total with nothing less than humanity's survival at stake, chemical and biological weapons are better left sheathed.

Stage IV: Orbital Insertion and Spacedrop

After a planet has been sufficiently softened for an assault, it should be carried out as swiftly and uniformly as possible. This is easier said than done.

Planetary assault will likely come in the form of two major troop deployments: orbital insertion, which implies a small-scale surgical deployment of troops somewhere on the surface, and spacedrop, which implies large-scale landings with massive numbers of fighters and equipment. In either case, landings must be conducted with absolute synchronicity to avoid losing the element of surprise, to effectively mass all fighting forces for assaulting their primary targets, and to hit all targets at more or less the same time to strategically paralyze the enemy for as long as possible. Once an assault begins, the time for diversionary tactics is past; a delay in dropping forces in one area of the planet will only delay assisting friendly forces in other sectors, and will not confuse any surviving defenders about the primary thrust direction of an assault.

As Earth's military history has shown, no pre-assault bombardment has ever satisfied the subsequent assault troops. Assault forces should expect hidden pockets of resistance, capable survivors, intact underground facilities, forces functioning at limited capacity rather than totally destroyed, civilian and non-combatant sneak attacks, and any manner of trouble which death from above could not solve. Assault forces would be wise to quickly and efficiently achieve their objectives, seek out the enemy political leadership, and compel a planetary surrender as fast as possible—for everyone's sake.

Assault troops should also be prepared to stay a while. This means carrying basing equipment, civil engineers, and infrastructure support needed to establish a planetary "beachhead," even if that beachhead lies in the enemy's capital. The going will not be easy, and will likely be long and hazardous; small-scale tactical decisions taken early in the assault could result in long-term consequences for the assault and resulting occupation.

Planetary Defense and Dealing with Local Resistance

All planetary invasions should begin with the expectation that a planetary occupation will follow. As with any occupation, resistance always rises to nip at the occupier's heels. Depending on the political goals of the war itself, occupiers should treat the conquered foe accordingly in order to keep the political leadership's options open. In a limited war, as in our example, excessive brutality realized on the populace will only hope to make post-war political normalization harder.

The forms planetary defense will take should not be surprising and will be roughly equivalent to familiar concepts. This is because planetary invasion shares universal concepts and truisms common to amphibious assault. Physical territory must always be physically defended; and the defenders themselves or some kind of technological avatar (automated war machines, surface to space defenses, and so on) must be physically present to do this. The first line of defense will always be left to the most professionally trained and best-equipped forces, if they are available. Further, if we continue with our amphibious assault analogue, the first line of defense is the "sea," or in planetary invasion, high or low orbit. In World War II, Imperial Japanese forces had only one strategy for defending against an amphibious landing: fire on and destroy the enemy at the "water's edge," namely, the beaches.[7] Beyond this, any strategy was wishful thinking; while the Navy Section of Imperial General Headquarters issued notifications to garrison commanders that every effort would be made to lure the enemy into the sea to destroy it, the fact was by 1944 there were no naval forces left to make that happen.[8] Consequently, Japanese defenders had to do the best they could in fortifying the islands, knowing full well they would be pitted against any amphibious invaders one-on-one and could count on no relief and no naval support. Japanese forces often prioritized fortifications on the beaches in line with their strategy, and in this regard planetary fortifications, even under orbital bombardment, make sense to construct and should be expected by any invader.[9] Given this, planetary invaders should expect a complex combination of near-orbit defenses "at the water's edge" as well as underground facilities relatively well-defended from orbital bombardment.

Invaders can expect the defenders to cling to whatever fortifications remain in semi-working order after orbital bombardment, construct new ones as combat proceeds, and field the remnants of their government-sponsored forces as best they can. At the same time, civil defense should also be expected, likely taking the form of aggressive paramilitary and guerrilla forces striking in small numbers at perceived tactical vulnerabilities. This latter point only reinforces the need to establish forward operating supply stations and mustering points in orbit when possible, and minimizing forward deploying invasion forces to the planet's surface unless they are actively conducting missions. Depriving defenders and resistance forces of the ability to launch their own counterattacks into orbit to harasses the assault forces should be a high priority; post-assault invasion forces will vulnerable while they are busy consolidating their positions in orbit and transitioning to occupying forces.

In general, dealing with planetary resistance from the local populace absolutely must be considered before the campaign gets underway. If a properly trained and prepared invasion force is to quickly make the jump from conqueror to occupier as seamlessly as possible, it must understand partisans and a local resistance will be a constant problem until the last human boot departs the planet. We are well-prepared for this eventuality from our own history, but we are doubtless unprepared for planetary-wide occupier duties. While this is mostly due to the sheer scale a planetary occupation would entail, there is no reason to believe we cannot successfully accomplish a planetary invasion and occupation.

In the case of a non-human sentient enemy, planetary annexation produces a special concern. If the reason for going to war was to obtain the planet for human colonization and habitation as part of our territory, human political leaders must think very hard about the cold logic that would entail. The enemy must be goaded and browbeat into granting political legitimacy to our demands, otherwise things will get messy very quickly. In the event negotiations go well and annexation proceeds, the enemy must be exhorted to completely withdraw their presence in a swift and orderly manner. There is one major reason for this: immediately following a war resulting in planetary annexation, peaceful coexistence between humans and the conquered species, combined with environmental and cultural factors, will be next to impossible. Not only this, but a population of the enemy remaining on the planet represents an unacceptable security risk due to espionage, sabotage, and so on. Such a population of enemy species remaining on the planet after human conquerors have won it in war and thereafter settled it can only result in any remaining enemy species becoming relegated to a second-class citizen status—whether formally or informally. A situation compounded by natural hate between different species which have just fought each other to the death cannot end well, especially during the relatively vulnerable early annexation period.

Local resistance, therefore, is best handled as follows: as quickly as possible, either repatriate the enemy population or exterminate them. The latter, while difficult from a human moral perspective, may be a necessary evil in the war's political circumstances. Unlike wars of old where a city's population could be despoiled of their possessions and then thrown from the city to scatter to the winds, planets, the islands amidst a lifeless void that they are, offer no such options to a conqueror. Coexistence is likely not possible due to security risks, racism, and species social and cultural incompatibility. Any assault force, and their political leadership, must approach planetary assault with their eyes wide open to such brutal and final possibilities.

As we can see, planetary invasion is the most challenging military operation which could be attempted in deep space warfare. Though it is neither easy, swift, nor kind, it may be the only way to achieve future political objectives in space. Whether the enemy planet is human or otherwise, scaling back from a planned planetary invasion can offer military planners a good perspective about the size, composition, and particulars of any military invasion in space. While it may seem distant now, physically taking far-flung space possessions from future enemies will remain one of the most dependable military operations available to future military strategists.

6

Economics of Interstellar and Interplanetary Warfare

Endless money forms the sinews of war.

—Cicero

In 1948, the People's Republic of North Korea was in the ascendant. Led by a charismatic and Soviet-backed leader, Kim Il Sung, the new state, with Soviet help, had evicted its previous Japanese imperial overlords and began the work of rebuilding its shattered economy. Kim, by accepting Communism as the basis for the state's new government (an offer he could not refuse from his new Soviet partners), co-opted the Soviet Union and China, thereby removing the immediate military threat from his northern border. North Korea's heavy industry and manufacturing sectors outclassed South Korea by a large margin, and while both states' GDPs were similar, North Korea held the clear advantage in military power with its large northern backers. Things were looking up for Kim in 1949 when the United States withdrew most of its troops from the peninsula as it continued to undergo major drawdowns following World War II, assuming a large land force would be unnecessary now that they possessed the power of the atom.

With a weak and divided South and a secure northern border, things were looking so good Kim sought Chinese and Soviet support for a military campaign to unify the Korean Peninsula by force. While he did receive this support, it was not enough to prevent a returning United States under a UN mandate from forcing a stalemate and returning Korean boundaries to the prewar 38th parallel.

In all this, the North's economy took a beating. As Gwang-Oon Kim puts it, "the 1950s was a trying time for the North Korean people. Having suffered an unparalleled level of devastation during the Korean War, they were left with the unenviable challenges of reconstruction from the ashes."[1] Besides the industrial bombing by American airpower, antiquated North

Korean industry and poor agricultural output was not up to the task of rebuilding the nation after two major wars in close succession. With nothing much on offer from his northern partners and unwilling to turn to the west to trade his way out of the problem, in 1955 Kim developed the distinctly North Korean economic mantra called *juche*. The word, which depending on the source means "subject," "main body," or "self-sufficiency," prescribes a laser-focus on economic ways for North Korea to produce everything it needs—alone.[2] This meant heavily industrializing agriculture to feed the population, make a profit, and generate enough internal economic activity to continue growth.

The policy was a disaster. The combination of closing off North Korea to the outside world and relying upon large-scale land collectivization with agricultural industrialization was the worst thing North Korea could do for its broken economy. As soon as North Korea adopted state-run farms with rural cooperatives, their entire farming population became state employees feeding a capital intensive and energy-dependent agricultural system.[3] North Korea's industrial sector became a slave to the agricultural sector, as it was the only possible sector where growth could happen, and there was little to no foreign trade to act as an economic lifeboat. A complete collapse was only a matter of time. In the 1990s, the inevitable happened. In 1991, North Korea was completely without trade partners, just in time to suffer a series of agricultural disasters from hail in 1994, floods in 1995 and 1996, a massive drought, then finally tidal waves in 1997.[4] Seed production and seed survival was not up to the challenge, as years of economic isolation had left North Korean state seed farms bereft of critical foreign agro-inputs necessary for successful hybridization and better plant survival.[5] Depending on the source, an estimated 200,000 to 3.5 million people died from famine in the mid–1990s.[6]

Juche continues to this day as official state policy, forever enshrined in stone as hallowed words spoken by the Dear Leader Kim Il-Sung, who still holds the office of Eternal President. This is in spite of the fact that Kim Jong Un, North Korea's current ruler, has stated his desire to bring economic development to North Korea. He has not yet attempted to reconcile these disparate policies, and likely never will. Abandoning the formal economic policy of the Eternal President is a difficult thing in the Democratic People's Republic of North Korea, and certainly impossible if the Kim family's legitimacy is at stake.

Juche is simply one of the more prominent examples of what happens when a society or civilization decides it can go it alone economically. Its implications and failures are clear: when states pursue economic isolation and non-dependency, they do so at severe risk. As a matter of fact, it is clear closed energy systems and economic isolation—voluntary or involuntary—are the two greatest problems facing interplanetary and interstellar civilizations.

The Planet as a Closed Energy System

In many ways, money is simply a representation of energy. Price is simply an agreement between two or more parties about who will expend what kind of and how much energy on a particular activity. Bills, coins, goods, rights, or anything tangible are simply tokens to prove this agreement has taken place. In this regard, interplanetary and interstellar civilizations must worry most about economics as they relate to the overall energy available within their domains.

The Kardashev Scale

One of the first thinkers to tackle energy usage by a civilization and relate it to its technological development was Soviet astrophysicist Nikolai Kardashev. In 1964, much like some American scientists Kardashev was employed looking for signs of extraterrestrial life.[7] During his work, he developed a scale by which he measured a species or a civilization's ability to utilize energy as well as its knowledge and relative technological development. There are three levels: Type 1, Type 2, and Type 3, all of which apply to a civilization as a whole.[8] A Type 1 civilization can harness and store all of the energy that reaches its home planet, and has reached an energy output of at least 10^{16} Watts.[9] A Type 2 civilization has not only utilized the totality of the energy which reaches it, but can also control an entire star itself. While this could mean many things, a Type 2 species could manipulate its star in whatever fashion it would like, which could include moving it or even completely encapsulating it within a device to harness it completely, known as a Dyson Sphere.[10] This type also corresponds roughly to an energy control and output of 10^{26} Watts. Kardashev's final type, a Type 3 civilization, is so advanced that its evolved or developed form would be completely different than its Type 1 forbearers. This species would be capable of galactic travel and total energy usage of whatever source it found. Additionally, Kardashev noted a Type 3 civilization must achieve an energy output of at least 10^{36} Watts.[11]

While Kardashev did not directly define a Type 0 species on his scale, by definition a civilization which cannot use all the energy which reaches its world, nor achieve a total energy output of at least 10^{16} Watts, is a Type 0.

Humanity is currently a Type 0 on this scale.[12] Our lowly Type 0 civilization is a lot like a pre-technological tribe. Inhabitants are capable of surviving and using what they find around them, but are subject to the privations and whims of nature because the total energy available for their use is only somewhat greater than what they need to survive. A planted field in a pre-technological settlement represents a significant input of that tribe's available energy; a single bad rain could flood this painstakingly planted field, which means that energy is wasted and people will die. Because a Tier 0 civilization cannot harness the total energy from its sun, its inhabitants must depend on energy created in relatively primitive ways, much like our tribe example. Indeed, the energy sources a Tier 0 civilization depends on are largely filtered or converted sunlight. Plant sources are simply sunlight which has been converted into solid matter via biochemical processes the plant uses to survive; animal sources are energy sources which are a step down this ladder as they must consume plant sources to create their energy. Fossil fuels are byproducts of centuries of carbohydrate (plant and animal remains) decomposition; even nuclear energy depends on radioactive elements formed originally by great pressures and clashes of cosmic and stellar energy, emerging from the proton-fusing blasts of the Big Bang and following few-billion-years of random energy fusions as matter expanded and spread throughout the galaxy.

Tier 0 limitations also means there is less total energy in the entire civilization as compared to Tier 1 and higher civilizations. Movements of energy are consequently more relatively harsh on its inhabitants since any energy reallocation represents a larger amount of actual energy available; a Tier 0 civilization's "total energy denominator" in a used versus available equation is an overall lower number as compared to a Tier 1 civilization. In our pre-technological tribe example, a single bad winter, a drought, a plague, a war, or any other catastrophic economic disaster can lurch the entire energy of a society in a few different directions, and the more energy that moves, the greater potential for poverty, loss, and even death. Even so, these situations do not totally destroy economic wealth; they simply redistribute it. Survivors of plagues profit greatly from snatching cheap land formerly occupied by their now-dead neighbors; weapon-makers do very well during wars. When economic systems become unbalanced, wealth is distributed haphazardly and often out of the control of its users.

Energy can be explained another way: it is the moving representation of material inequality. In a closed system, every joule of energy either belongs to someone, is in storage, cannot yet be utilized, or is waste. Consider another example: an international businessman getting a raise. For ease of understanding, in this example this businessman works in the fishing industry. His raise happened because his business was doing better and became more profitable; this meant, by definition, his fishing industry was outperforming other fishing industries in a closed system (in this case, the global economy). His raise represents a redirection of energy, and that energy has to come from somewhere else in a closed system. In a world with a saturated fishing industry market, one or more of his rival fishing industries by definition must have been less profitable, and may or may not have had to make difficult economic choices because the energy available to them decreased in quantity. If carried on long enough, the losing fishing industry or industries will have to dismiss employees or otherwise shrink in size in order to ensure the survival of their business as a whole. Furthermore, less energy in the form of fish—a commodity which can also be converted to biological energy—will be available to their customers. These fired employees, who were not necessarily poor workers and could even have been more skilled workers than the promoted businessman, are victims of energy inequality found inherently in a closed system. In short, small effects in one part of a closed system, no matter how large that system is (in this case the globe), have localized economic effects somewhere else as well as systemic effects on the whole system.

This changes in a Tier 1 civilization, but only slightly. The chief advantage of a Tier 1 civilization is the *total quantity* of energy is greater, making day to day existence much more pleasing as basic commodities are more available and basic inequalities are removed. Energy is always converted by a species into something useful for that species. Whether it is money, food, labor, tools, or technology, energy resources exist to be consumed by those that find them. While energy distribution will always be uneven, one of the best ways to even out energy inequality is to pour high quantities of energy into a closed system. The entire energy of a sun could do just that. But the ability to use the sun's energy output is only effective in removing energy inequality because it is so much greater than the energy currently available on Earth, as it is a closed system. In other words, because humanity cannot yet utilize the sun's energy in its entirety, it must still to some degree rely on energy sources here on Earth, which are spotty and unevenly distributed. By increasing the total overall energy to obscene quantities, the ubiquitous availability of energy in whatever form it takes is vastly preferable to the energy inequality we have now. Even if most of the energy from the sun is wasted, the titanic quantities—10^{16} Watts by Kardashev's reckoning—will drown out the distribution inequality. It would be like solving Earth's current starvation and food distribution problem by dumping foodstuffs over every square kilometer of Earth's surface from the air; while colossally wasteful, there would no longer be a distribution problem, and no starvation.

Deep Space Economic Activity

Luckily, it is not required to achieve the distant Tier 1 status before heading out into the stars. Indeed, it is safe to say interstellar economic activity will first serve the needs of those that venture out first to claim its resources. In fact, one of the primary impetuses to future solar system exploration will be to find things which solve our problems here at home. This requires a savvy combination of discovery, technical skill, risk, and luck.

Because this book is about deep space warfare, it will not discuss the necessary technological steps or the technical adjustments required to make the first leap towards establishing planetary settlements, intrasolar mining operations, or interplanetary economic resources. Once these things are established, their relationships become clearer, as do these relationships' consequences for security and war.

Interplanetary Economic Dependency

Among our species' first goals will be using resources found in space to solve problems here on Earth. Given what we know about the limits of a closed economic system, this makes sense; injecting energy from outside the system into our planet-locked economy can resolve a host of tenacious issues, including food distribution, starvation, energy shortages, and overall wealth disparities.

However, this also comes with the risk of economic dependency. Imagine, for example, if a settlement was established on Mars. The chief purpose of this settlement is to produce foodstuffs to feed Earth's growing population. With a little technological imagination, this is not too far-fetched. A planet-wide farm destined for interplanetary exports is a noble effort; it makes sense to feed people who are starving, and if we can make our lives better, we should. But once this farm succeeds in its purpose, the energy it introduces into Earth in the form of foodstuffs will create a dependency. The people who once went without when there was no Mars farm will now depend on it for their survival. Similarly, since food availability is one of the main factors in population growth, the many millions sure to be born succored by the Mars cornucopia will also depend on it for their lives, as will their descendants. In short, the transportation of foodstuffs to Earth from Mars now becomes a literal and figurative lifeline, and therefore a species security issue. Defending it is critical—and to any adversary, attacking it is tempting and very effective.

Beyond food, an energy dependency potential quickly becomes self-evident. If NASA is correct in its estimations that there are 700 quintillion (yes, *quintillion*) dollars' worth of materials to be found in our solar system's asteroid belt alone, future miners and corporations who introduce these materials into Earth's economy will also unintentionally create an industrial and economic dependency on them here on Earth.[13] In fact, "materials" or "food" can be replaced with any commodity, to include population, which of course serve in militaries (assuming future policies seek human additions to future automated forces). In our above settlement example, distant settlements could one day provide priceless commodities for Earth. The lines of communication and transport between these worlds instantly become critical security issues.

The Concentric Economic Exploitation Model

For many practical reasons, it is clear humanity will expand its economic activity in space concentrically outward, starting from Earth. This will happen due to two main factors: the formidable distances involved in reaching ever-farther outward in space and the logistical difficulty and expense which comes with it, and because of the need for economic access to each preceding layer before expanding to the next. Both these factors have significant military implications, as we shall soon discuss.

Even now, we can clearly see the concentric levels of this future economic space exploitation. Extending out from Earth all the way to the asteroid belt, these "layers" of potential economic activity hold resources and benefits awaiting exploitation by the first intrepid explorers who can get there and claim them. They are as follows, in order of closest proximity to farthest.

- **From lower Earth orbit to Geosynchronous (GEO) orbit ranges.** While this domain is currently the exclusive purview of satellites, it will not remain so forever. GEO ranges are seductive economically for their easy access to solar energy and relatively close range to Earth's surface, thereby maximizing the efficiency of future solar power plants as they will be able to beam their energy to surface facilities for Earth use, with less energy becoming lost in transit compared to solar plants farther away. Further, GEO orbit ranges are perfect for future space tourists, and ideal locations for hotels to house these tourists prior to the next leg of their journey. Hotels in GEO orbit could even offer a particular gimmick which entices tourists—a constant orbit above a particular continent or geographical feature could have a great draw for those on Earth interested in that kind of opportunity. Finally, GEO orbit is best suited to research stations as well, especially if that research pertains to the below geography, the weather patterns of a particular continent, thermal applications, or anything a research entity or corporation may find useful from constant observation.

- **Near Earth Objects (NEOs).** A certain type of stellar body classified as NEOs range between the vast space between Earth and the moon. While it seems close from a cosmic point of view, there are nearly 240,000 miles between here and there, miles which are filled with both comets and asteroids. According to NASA, objects which pass between Earth's immediate proximity and 1.3 astronomical units (1 au = appx 93M miles) are classified as NEOs, and the vast majority of these are asteroids.[14] As such, NEOs form the next rung on the great economic stepladder reaching towards the outer solar system. These asteroids and comets contain water, metals, volatile chemicals, and likely a variety of as yet undiscovered alloys and compounds. In fact, by some estimates the minerals and substances found in NEOs could sustain a population up to 150 billion people, and the water alone could sustain 400 billion.[15] Naturally, these populations could never fit on Earth's surface, but the fact that such economic potential energy could exist in minor objects which spin their way past our planet is astounding.
- **Mars.** The first major planet-sized human settlement will probably be on Mars. For a variety of practical reasons, Mars has already invaded the imaginations of science fiction writers, space enthusiasts, and practical space policymakers alike. Mars' day and night cycle is almost identical to Earth's (23 hours, 59 minutes, 4s for Earth, 24 hours, 40 minutes for Mars).[16] If one could someday step outside onto the Martian landscape and enjoy the scenery, they would find seasonal cycles similar to Earth since both planets share similar axial tilts (23° for Earth, 25.19° for Mars), though no breathable air nor changing leaves to enjoy.[17] What Mars does have, though, is a great deal of frozen water at its polar caps, and likely a large amount below its surface as well. Where there is water, there is energy; and besides water, Mars possesses a large amount of CO^2 in the atmosphere and minerals below the surface. Mars' red appearance is caused by iron oxide, common rust, which can be chemically reduced to iron ore and oxygen. While the first harvests of these minerals and materials will likely be used to support Earth's first settlers, once they achieve a basic energy subsistence, there is no reason to believe these minerals could not benefit Earth and humanity as a whole.
- **Asteroid Belt.** Once we are able to mine the asteroid belt, located between Mars and Jupiter, solar system economics will really take off. Consisting of over 790,000 individual rocky bodies floating silently in the blackness, the mineral wealth located in these objects is thought to be stupendous.[18] The mineral content will contain trace to large amounts of just about every mineral and substance we have ever discovered, and perhaps even more we

have not. This is because the asteroids in the belt are simply leftovers from the materials which originally formed our solar system, and as such contain approximately the same variety and concentration of the same material composition of the solar system's planets. While these asteroids were never large enough to be considered planets, their gravitational behavior is similarly predictable; they travel via measurable orbits based on the gravitational pull of their nearest massive bodies, which includes Jupiter and the sun. This means these objects can be mapped, scouted, and predictably mined. Its contents are thought to be able to support the mineral and economic needs of 5,000,000 *billion* people, wherever those people happen to live; the water content of all asteroids in the belt could sustain up to 30,000,000 *billion* people.[19] While these estimates are conservative, it is important to note that they are probably not that far off; it is difficult to truly comprehend the mineral wealth awaiting those that can successfully reach and exploit this floating mother lode.

- **Beyond.** Clearly, there is more to see after the asteroid belt, and humanity will not stop. Outside the confines of our solar system, an incredible variety of stellar phenomena await exploration and exploitation. There is no telling what we will find economically useful farther into space, and the utility and composition of many different stellar bodies have yet to be explored.

Strategic Resources in Space

King Water: Nectar of the Gods, Giver of Thrust, Coaling Station of the Future

Where there is water, there is an economy. Where there is an economy, there is war. Water is the most important and versatile economic commodity in the solar system, but it is nowhere near the most abundant. Like carbon, water's unique chemical properties and its chemical utility make it an indispensable necessity for a spacefaring civilization. Where carbon's tetrahedral structure makes it the most likely building block for biological life, water possesses properties which are essential to both organic and inorganic chemistry. Water easily ionizes, making it an excellent solvent and foundation for millions of different chemical compounds. It is relatively easy to transport in solid, liquid, or gas form, none of which are especially reactive, volatile, or explosive, something which will certainly come in handy during long voyages across the vast reaches of cold space. Its ability to react with and form multiple chemical compounds, including both organic substances consumable by humans and inorganic fuel sources like rocket fuel, make it the most indispensable substance in the laboratory.

From water, humans create their most essential spacefaring support materials. This includes breathable air on spacecraft and far-flung human settlements, drinking water, rocket fuel, and nuclear thermal propellants. Water satisfies human hygiene needs and enables humans to grow and raise the foodstuffs needed for life. Water's atomic components can be separated into hydrogen and oxygen, which can then be used to create liquid oxygen (LOX) and hydrogen, both critical fuels for spaceflight. This atomic separation will likely be the future source of breathable oxygen on future spacecraft and in human settlements, especially places with no atmosphere and a reliable source of nearby water, like the moon.

The implications of nearby water in the form of ice on a planet or planetoid with no atmosphere have been known for some time. Such places, like the moon, are excellent fuel stations for spacecraft on their journeys through space. The lack of an atmosphere at such a fueling station means no vessel need strain itself in achieving enough speed to escape it, and thus avoids the risk that comes with being too slow trying to pierce an atmosphere. Water, meanwhile, cares little for the presence or lack of an atmosphere; cold water is simply ice, and therefore becomes arguably easier to handle in solid form by human refuellers.

The combination of a large source of water, a nearby fueling station that can separate water's molecules into its component oxygen and hydrogen, and the lack of an atmosphere are the building blocks of these future "space coaling stations," and powerful incentives to economic activity in their own right. Indeed, this combination could one day exist at many different locations in the solar system, allowing human space travelers to "stair step" their way to the asteroid belt by hitting critical fuel stations along the way. Mars certainly qualifies as one such potential fueling station, as does its moon Phobos. In fact, the list of potential locations where water could be found in liquid, solid, or vapor form is striking; Saturn's Enceladus, Jupiter's Ganymede and Europa, and the NEOs which fill the void between Earth and the asteroid belt are all excellent contenders for these modern-day space coaling stations.[20]

Rare Earth Elements

Besides water, other key resources in space include things humanity values: rare earth elements and other minerals. Substances found and fought over here on Earth in relatively small quantities likely exist in unbelievable abundance on NEOs and in the asteroid belt. One such item, called ilmenite, consists of titanium, oxygen, and iron, and bears the chemical formula $FeTiO_3$. This compound is the main source of titanium here on Earth, a metal whose alloys are critical to the aerospace industry and to producing spacecraft and industrial equipment capable of standing up to the stresses found in space. This compound is thought to exist in titanic quantities in NEOs, the asteroid belt, and even just below the lunar surface.[21] If even a relatively small spacefaring cache of ilmenite can be secured, the owner will likely have a source of titanium which dwarfs the small concentrations found here on Earth. Besides ilmenite, a variety of rare earth elements, alkalis, and alkaline earth metals are thought to exist in massive quantities silently below the surface of these planetary objects. The military implications of these facts are obvious.

Oxygen and Other Gases

While gases do not immediately come to mind as being militarily useful compared with images of mass drivers and missiles, in fact gases are critically important to economic and military activity in space. As mentioned above, gases are the principal fuel source for rockets, especially hydrogen. Oxygen, found mostly in compound form bonded to iron, titanium, hydrogen (as water), and many other chemicals, is still a relatively common element in space and absolutely essential for human life support and propulsion. Once chemically freed from its parent compound, free and gaseous oxygen can then be combined with nitrogen to craft the breathable atmosphere needed for spacecraft and human settlements.

Further, the gasses waiting to be investigated and exploited in nebulae and our own solar system planets (like Jupiter and Saturn) will certainly one day have economic and military applications, though we can only speculate now as to what that could be. These gases could one day become an economic mining source in their own right, if it suddenly becomes energy-efficient to separate the gases found naturally in the solar system's outer planets into their chemical components.

Competition vs. Deprivation: Plenty to Go Around for All?

Even the vastness of space may not be enough to deter conflict between flag-waving territories and political units. Space's material richness will naturally push political organizations to value the same areas and stellar phenomena, since all human organizations are run by humans, and presumably value the same resources. Such resources could include energy and mineral wealth, unexplored scientific phenomena, militarily strategic locations, or even habitable planets. Given the difficulties in claiming, exploiting, and travelling to and from these resources, it is therefore reasonable to assume humans will fight each other over these resources the same way we have always done. It is always cheaper and more efficient to take what your neighbor has made, especially if those things happen to be closer than ones which are unclaimed but ready for the taking.

At first, the vastness of space may encourage strategists to think there is plenty for all to go around; enough mineral wealth, unexplored territory, and gaseous phenomena to exploit for every player. Soon, though, it becomes clear that not all such resources are created equally. Some will have less or more depending on what speculators actually can find, and what current technology will allow them to extract. In all cases, the expense associated with distance to and from space resources will encourage rivals to creatively appropriate others' hard work which may be more advantageously located. Protecting these claims will be a major role for future space forces.

Farming Farmland: Agricultural Benefits of Interplanetary Settlements

Being able to militarily claim and protect a system and its planets has clear economic benefits, which then feed back into military power. For example, with extraterrestrial settlements, we could theoretically farm farmland. What does this mean? Simply put, dirt—of any kind, from any planet—can be converted to soil with the simple addition of water, chemical minerals (especially nitrogen and phosphates), bacteria, and light, all in correct proportions. There is plenty of water ice to be mined in the galaxy, and can be created in a lab in any case. Nitrogen and other minerals can be fixed from all sorts of stellar phenomena or chemically separated from minerals found in NEOs and asteroids. Bacteria are self-replicating once placed in their environment and given a modicum of food sources. The point is this: in creating arable extraterrestrial soil, a minimal amount of energy and matter originating from Earth can go a long way towards enhancing Earthbound soil and depleted fields across the planet.

We already know soil quality upon Earth varies greatly from place to place. But consider comparing, for example, the absolute worst soil available for growing crops on Earth with that of dust from Mars; there will be a difference, but not a great one. By introducing dirt from *outside* Earth's planetary ecosystem, farmland can be *created* from previously dead and useless soil. After ensuring no extraterrestrial bacterial contamination, there only remains the problem of where to put the newly created farmland. Depleted soil on Earth, whose rejuvenating mechanisms either take too long (like growing forests on top of depleted soil) or have been redirected by humans for other priorities, is a good candidate. Orbiting hydroponics facilities are also a potential option, though expensive, and theoretically limited only by the size of the facilities. Viewed from this perspective, near orbit begins to resemble quite a prairie high above the planet. Settlements, especially struggling ones or those lacking in soil fertility, are another good choice. In any case, until this dirt becomes available in massive quantities, Earth is stuck with the matter it has locked within its atmosphere.

In breaking the millennia-old pursuit of planetary-wide subsistence agriculture through the introduction of this extraterrestrial farmland, Earth's economy is doubtless in for a shock—but it could also be the first step to truly eradicating planet-wide poverty. Since at its core an economy is based on agriculture, this idea obviously conjures up concerns that the influx of incredulous amounts of grain and produce from settlements and orbiting farming stations could completely destabilize Earth's economy. After all, cheap Roman provincial grain savaged the Eternal City's economy, disenfranchising Latin freeholding farmers and saddling the city with economic instability and a public dole second to none.[22] Granted, this idea has its dangerous aspects. Some population economists would likely point out that it is precisely *because* Earth has limited room and limited farmland that our population is kept in check. Other economists would no doubt retort that as food becomes more reliably available and as a population eats its way up the food chain by including more protein in their diet, population growth tends to slow and stall. Still others point to distribution woes which would accompany our current inadequate food distribution systems. Moreover, if Earth becomes dependent upon off-world foodstuffs to maintain its population, similar to many island nations today, access to this food then becomes an obvious target for enemies in future warfare, and forces the planet's population to hitch its survival to the technology needed to transport these life-sustaining goods. However, if this chance could truly end world hunger, there are millions of luckless and malnourished humans across the globe ready to give it a try.

While agriculture is a key winner from extraterrestrial settlement, there are a myriad of other human social benefits which result from peaceful extraterrestrial settlements. The scientific discoveries waiting to be made are by themselves are enough to justify any settlement attempt. However, these benefits will be left to others to discuss in other volumes; for now, we must return to military matters.

Mahan's Ghost: Economic Warfare in Deep Space

There is an old saying among pilots: "airspeed is life; altitude is life insurance." The faster an airplane goes, the more lift it generates, and the more guaranteed it is to stay in the air. If a pilot somehow manages to get too slow, the altitude in between the airplane and the ground will give the pilot time and energy to get that speed back. In space, this expression can be harnessed again, but more resembles the following: "energy is life; energy storage is life insurance." In space, it is energy which powers spacecraft, energizes life support systems, and makes transit through its inky blackness possible. Empty space is a near-zero energy vacuum, which silently and continuously drains any passing body of its precious energy for its limitless entropic appetite. Space bears no malice in this regard; it is simply following the second law of thermodynamics, which states the entropy of any isolated system always increases.

Energy storage grants the necessary margins for error when things go wrong; when a particular system on a spacecraft or shelter breaks down, or when a force must flee from battle, or a when navigator miscalculates a course and a vessel suddenly finds itself millions of kilometers from a resupply point.

These energy problems have been known for centuries via our economic activity on the oceans. As we have already discussed, space's natural environment lends itself to a maritime model, in both military and naval aspects, and one of the first to lay out these maritime energy issues as matters of national concern was Captain Alfred Thayer Mahan, who was mentioned earlier. Now that we have discussed what is economically at stake in space, the next natural question to ask is how these things can be safeguarded and exploited with relatively little harassment by other interested parties.

Mahan understood modern nations depend on the sea road, and the trade and material wealth that comes with it. In his landmark work *The Influence of Sea Power Upon History*, he divided a nation's maritime wealth into three distinct categories: production, or how nations make the commodities traded via the sea; shipping, or how these commodities are exchanged; and colonies, which multiply production capabilities and safeguard a nation's supply by providing multiple production sources.[23] He says:

> In these three things—production, with the necessity of exchanging products, shipping, whereby the exchange is carried on, and colonies, which facilitate and enlarge the operations of shipping and tend to protect it by multiplying points of safety—is to be found the key to much of the history, as well as the policy, of nations bordering upon the sea.... This order is that of actual relative importance to the nation of the three elements—commercial, political, military.[24]

Mahan explains a nation's wealth is best secured by a strong navy, which in turn can both protect a state's disparate economic interests, located on various continents across many different seas, and can hold an adversary's similar industrial wealth at risk. To do this, Mahan concludes a navy's first major military purpose should be to destroy an enemy's maritime commerce.[25] Calling upon historical examples of England's rise to maritime supremacy after it both destroyed its rival Dutch vessels in battle and strangled its maritime trade, Mahan makes a strong case for commerce destruction, and friendly commerce protection, as critical roles for any navy. He points out that commerce-conducting vessels "having little power to defend themselves, need a refuge or point of support near at hand; which will be found either in certain parts of the sea controlled by the fighting ships of their country, or in friendly harbors."[26]

When it comes to engaging an enemy battle fleet, Mahan calls for gathering and concentrating a fleet in force, with the purpose of destroying an enemy fleet utterly—an aim which would have made his hero Nelson smile. This, of course, is easier said than done, especially since this kind of strategy both takes away from the navy's commerce destruction and protection roles, and is made more difficult since an enemy will ostensibly be trying to do the same. Furthermore, critics of Mahan's approach note how difficult it is to concentrate naval forces, as each singular vessel is capable of different cruise speeds, not to mention the communication problems associated with contacting and coordinating many different far-flung ships scattered across the globe. Mahan's age was switching from sail to steam; when he wrote, the age of the battleship was still in full swing, and "wireless" or beyond-line-of-sight electronic communication was a brand-new technology. Now, with instantaneous and encrypted over-the-horizon reliable wireless communication, naval forces are better prepared than ever to respond to an incident at sea or concentrate at a particular location for battle. While space will certainly challenge a naval force's ability to concentrate due to its challenging distances, it is reasonable to believe the technology to communicate across vast distances will be available as soon as the first battle fleet in space is built. The coordination this technology will provide is only the first, but most necessary, step towards achieving Mahanian—perhaps Nelsonian—naval dominance in space.

Mahan's Three Economic Levels in Space

Despite Mahan's theoretical shortcomings, his three-level approach to maritime economic activity is clearly applicable to economic activity in space. Future spaceborne military activity should focus on these three aspects to maximize military impact in space, especially since we already know a battle fleet's protection and presence is sure to be desired all at once by every friendly economic base and settlement once war breaks out.

First, Mahan notes the facilities and centers of production are central to maritime warfare.[27] Production in space is represented by planets, future orbital bases, construction facilities, and any other settlements or organizations which produce war material for their political owner. This also means the agricultural backbone which powers these industrial economic appendages. For now, production is limited to surface-based Earth facilities only; but it would be foolish to assume this will remain so forever. As we discussed earlier, the moon and other planetary bodies which contain local sources of water will function as future spacecraft versions of coaling stations, which means they will also come to function as full-time service stations as well. This will lead to repair facilities, local economies and facilities to support the particular needs of shipping and spacegoing vessels and their crews, and therefore eventually to spaceports and full-blown parts and spacecraft production facilities.

Further, the efficiency which comes with constructing spacecraft and space-centric machines in a zero-gravity environment is simply too seductive to ignore for long. In zero-g, it does not matter how massive in size a spacecraft component or construction materials become; since spacecraft undergoing orbital construction are already free of an atmosphere, there is no need to worry about the thrust-to-weight ratio needed to escape an atmosphere, which we must worry about now for craft constructed and launched on Earth's surface. In addition, moving the components and fuel about in space is as simple as giving it a little nudge; when everything is weightless, entire components and massive hulls of future spacecraft can be assembled with relatively little force. The process can also be better automated; free-floating construction robots will function brilliantly in space, especially since construction can be continued round-the-clock with little regard for robotic work-rest cycles. The lack of seasons, weather, and catastrophic phenomena high above earth's troposphere means construction can continue all year round with no appreciable impact on work schedules due to changing weather and natural disasters. While the temperature will always be very cold, specialized tools and construction techniques for zero-g, near-absolute-zero construction only need be pioneered before Earth orbit becomes the construction area of choice for spaceborne machines and spacecraft. Production, then, will include orbital facilities as well as those on the surface.

Second, shipping is clearly analogous to spacegoing transport and mining vessels sure to appear once we begin exploiting resources in space. Similar to maritime trade, at first these ships will be relatively rare, slow, and consequently relatively higher in importance to the economies which depend on them. It was not long ago that terrestrial sailing ships were welcomed into ports bearing the only cargo that nation would receive from its particular source for that year; oranges from the New World, silks and spices from Asia, ice from the polar caps. These vessels were anxiously anticipated and often brought the preponderance of profit for any particular trading company. Because space transport vessels will be expensive, hard to crew, take a very long time to conduct their mission, and are vulnerable every second they spend in space, their importance to their economies and their companies will be correspondingly high; as will be the prices of their goods.

This also means they become excellent targets in war. A smart belligerent will not destroy these vessels, but rather capture them; the resources on board will likely be just as important to one side in a conflict as to the other. Further, the relative dearth of these vessels and expense accrued during their construction means capturing them and re-flagging them is more economically sound than simply obliterating them along with their cargo. Finally, because of the vast distances of space, dedicating military assets to defend these important transport vessels largely depends on where they are and not on how many there are. This means defending transport ships via convoys is conceivably much more efficient for their owner than trusting luck to see how many can get through an enemy presence. Likely, future space battles for shipping routes will see a return in force of the naval convoy, especially as economic activity advances to the point where states become more reliant on resources found in space and the concomitant need to safeguard space shipping increases significantly.

Colonies, of course, are economic outposts from mining stations to settlements and everything between. According to Brent Ziarnick, colonies are any place in space which actively utilizes, are products of, and represent the projected space power of its owner.[28] Additionally, colonies serve to enhance and project this power further into space. For instance, Ziarnick points out today's modern-day space colonies are satellites; while they are not the pure products of space power per se, they enhance it by acting as platforms to carry out space business, like communication, military intelligence gathering, and commerce.[29] Colonies are any asset placed in space to manage, change, or utilize space power; as such, their definition is broad. Put another way, colonies are the living representations of the political units seeking to project and use space power.

It is widely believed colonies will eventually expand to encompass inhabited human settlements, wherever they happen to be. Industrial mining, resource gathering activities and installations, and semi-permanent and permanent presences in space all fall under Mahan's definition of colonies. As such, their significance to future deep space warfare is profound.

The Importance of Access

If one believes Mahan's way of thinking about space, which is a compelling analogy, then access becomes the most immediate problem of space resources—and the most immediate target of potential adversaries. Mahan's ideas about economic warfare on the high seas clearly centers around denying the enemy its vital shipping lanes and access to needed industrial materials via its shipping fleet. Better still, denying an enemy safe havens and port access can only benefit a nation's space ambitions. If our concentric model of economic development in space holds true, then any military force or economic vessel conducting activity in the outer parts of the ring, like the asteroid belt, must necessarily have access to the inner parts, like GEO orbit, if it is to successfully deliver its cargo or provide its economic benefit to its state on Earth. This means military access to each layer of economic activity is essential to utilizing that layer, and the layers in between it and Earth; if access ever becomes denied or cut off by an adversary, the economic benefits of controlling an outer economic layer can be neutralized. In short, Mahan predicts a model of warfare which consists of states seeking more maritime benefits for themselves, and depriving rivals of similar maritime benefits. This translates very succinctly to space warfare, and is as good a place as any to begin thinking about deep space military strategy.

Critics of Mahan's approach note how not every nation is as dependent on the sea as others, and thus his maritime strategy may be futile. Landlocked Mongolia, for example, depends more on land and airborne trade than maritime activity; certain South American countries have almost no naval power, despite their access to the sea, due to their relatively smaller dependence on oceanic trade and industry. But this criticism will not hold in space for one simple reason: in space, *every economic possession depends on maritime space access*. Planets, planetoids, and locations or objects which humans could settle or possess economic interests in are 100 percent surrounded by space. This means every single asset, facility, and location, including Earth, is completely dependent upon its access to the "sea"—to space—for its well-being and its survival. It is clear Earth and its population will one day come to depend on resources gathered from outside the planet; rare minerals and water from NEOs, fuel for spacecraft from the moon, perhaps even one day Earth will become dependent for survival on agricultural imports from orbiting farms or other planetary settlements. Because all of these locations—Mahan's colonies—will be located somewhere in space, every single one will need safe and reliable maritime access. There is no such thing as "landlocked" in space.

A Spacegoing Merchant Marine

Given this, and if economic resources in space are even minutely as valuable as we think, and if military doctrine directs defending production, shipping, and colonies, it follows that creating a space merchant marine is the best way to defend the shipping routes and disparate economic resources we value.

Here on Earth, merchant marines have often found themselves playing second fiddle to their respective Armed Services, a fact which is also true for the United States. Nevertheless, the Merchant Marine and Navy are not only separate entities with different missions, but are interdependent and essential for a sound maritime economic policy. One of the most prominent naval thinkers to comment on the importance of the Merchant Marine in the United States was Rear Admiral Stephen Luce, founder of the U.S. Naval War College. As the 20th century dawned, the United States was rapidly realizing its naval and economic power, and naval officers were on the forefront of this newfound American global vigor and power projection. In 1903, Luce gave a famous address at the U.S. Naval War College which discussed the Merchant Marine and its importance:

> An intelligent study of naval policy must necessarily include our shipping interests. The military marine and the mercantile marine are interdependent. The navy, while policing the sea, protects our foreign commerce, and in time of war, finds there its greatest reserves. It was once observed that we had "clipped the wings" of commerce and driven our carrying trade to foreign bottoms. The same is practically true today. Thus we are not only contributing indirectly to the support of foreign navies, which may some day be opposed to our own; but we are depriving ourselves of what would prove, in time of war, an auxiliary of incalculable value. The remedy for this deplorable state of affairs must, necessarily be left to the wisdom of Congress. But the navy, with no other interest in the question save that dictated by the highest sense of patriotism, discharges an imperative duty, in urging as a military necessity, the re-habilitation of our mercantile marine.[30]

One of Luce's main points was that while having a navy is essential to safeguarding national political interests, having a solid, dependable, nationally-flagged shipping fleet—a merchant marine—is just as important to protecting national economic interests. It also functions as a countermeasure to finding oneself losing access to another foreign nation's protection for one's shipping fleet due to war. In Luce's day, Great Britain provided the global naval presence which safeguarded international shipping lanes, which included U.S. merchant vessels; if Great Britain ever went to war, Luce's concern was that protection would suddenly disappear or become re-prioritized to exclusive British interests during the conflict. If the United States and Great Britain ever went to war, the protection provided by British ships would suddenly become a vital national threat.

Scarcely eleven years later, Luce's prediction came true: trans-Atlantic American commerce found itself smothered beneath German U-boat attacks targeting British-flagged vessels, which included vessels assigned to protect shipping. Soon thereafter, Germany's unrestricted submarine warfare began decimating all vessels in the Atlantic, including non-belligerent American ones. The lessons are clear: whether or not one is at war with a power capable of destroying one's shipping vessels is irrelevant to the need for a Merchant Marine. This lesson must be applied to future spacegoing commerce if a state expects to protect its economic advantage from space resources.

Space Economic Warfare

It seems, then, a spacefaring navy's military goals are clear. Ideally, space navies can conduct economic warfare against their foes by seizing control of areas of space, safeguarding friendly commerce and shipping vessels, destroying or grabbing enemy commercial ships, and threatening or controlling any possible friendly port an enemy or other naval force could use for their own gain. Given the distances involved in space, we already know naval vessels of all kinds will be in high demand and short supply. If history is any guide, economic activity will consistently outpace the military presence required to defend it; there is always some entrepreneur or greedy corporation who disregards material risks to chase the siren's song of "get there first." Humanity should become comfortable with the likelihood a military force will follow economic pioneers as they expand into space. Like old U.S. Army forts established in the American West to protect settlers as they moved into "Indian country," maritime forces should be expected to appear when they are needed, and not before. The state, like a sun, has only so much energy to expend, and only the truly economically strategic areas can expect a robust spaceborne military presence.

The implications of Mahan's predictions on space combat are also clear. Should war break out someday in space, a navy's first target ought to be the commercial capabilities of an enemy spacefaring state. Clearly, the greater an enemy's reliance on its production, shipping, and colonies, the more decisive and strategically significant a maritime attack on these things will be. As of this writing, since no state is truly dependent for its survival on the space version of Mahan's three economic levels, it is difficult to claim this kind of strategy would be successful during a war in the near future. Nevertheless, space navies must be built with an eye towards these objectives if they are to someday be successful in future space warfare.

As we have seen, economic activity is a major reason to go to space, but is not without its dangers. This chapter examined the various resources awaiting intrepid states between here and the asteroid belt, their economic and military significance, and how maritime forces should posture themselves to defend their economic interests and to menace their enemy's. While there are significant technological developments that need to be made prior to engaging in economic activity similar to Earth activity, we can be confident this will be done not by accident but by those with the desire to reap the rewards. If a state is truly interested in seizing the economic boons found in space, it must also be ready to defend them.

7

Dealing with Non-Human Cultures

> Such then is the human condition, that to wish greatness for one's country is to wish harm to one's neighbors.
>
> —Voltaire

For the foreseeable future, humanity's most likely adversary in the stars is humanity. It does not take an overactive imagination to predict how terrestrial concepts of property, security, nationalism, and others will spill over into stellar domains where they will play out yet again on a grander, wider, and more distant stage. A land and resource grab akin to the discovery of the New World could easily occur once again in deep space, with stories of gold and riches giving way to tales of energy and strategic resources—perhaps even a new Earth-like world on which to settle. This latter possibility, especially, could easily fuel ambitious and jealous nations to contrive their new world into producing a "nationally pure" or even "racially pure" planet. By seizing a world by force and imposing one of our more nefarious ideologies onto its colonization, a spaceborne colony could quickly become a fascist's or tyrant's dream.

A worst-case scenario, to be sure, but unfortunately familiar. Despite these prominent risks, the possibility should be examined that humanity's spaceborne adversaries will not only be terrestrial, but could also be another sentient but non-human race of organisms. The sheer variety in which non-human biological organisms and societies could occur makes it impossible to predict what level of danger and competition will result from our two (or more) species encountering each other. Therefore, the most prudent practice is to assume the worst and, for military purposes, concentrate on potential conflict resulting from non-human cultural interaction.

The Problems of Communication

Inter-species communication deserves its own chapter, perhaps its own book, mainly because it is likely to be the only thing between peaceful conflict resolution and war.

The subject of communication is very well studied, understood, and analyzed here on Earth. Many make their living peddling their particular brand of enlightened and effective communication practices, and still more industriously study the very foundation of the concept itself: from humanity's simplest grunts and word utterances to the complicated tapestries of written and verbal language families. In any case, the method or mode of communication eventually chosen to communicate with a non-human culture is not as important to this book, as it is assumed upon first contact with a sentient culture we will not be able to immediately communicate in any case. But once the keys to our preferred modes of communication are discovered and communication does become possible, the real trouble can begin.

Communication is inherently difficult due to its fundamental biological flaws. Mixed with a healthy dose of mistrust, misunderstanding, and ambiguity, communications with non-human cultures become even more difficult. Even here on Earth, those of us who speak multiple languages regularly point out "untranslatable" phrases and concepts, hidden meanings, and symbolic language regularly utilized by many Earth cultures. Imagining effective non-human communication becomes harder when one acknowledges our entire knowledge of language is currently confined to one species, and this single species by and large has the same basic cultural concerns and a great deal of common problems to solve (family, food gathering, security, birth, death, and so on). Yet somehow it remains true there are some concepts which are either too obscure or too culturally particular to communicate without experiencing firsthand. This presents a great deal of problems to the exolinguist attempting to decipher an alien culture's language.

This brings us to the crux of the problem of intercultural communication: errors. In the best of cases, with each side of an intercultural or interstellar communication possessing the best of intentions, and genuinely attempts to really and truly communicate with the other, errors are bound to occur. Unlike a tourist in a foreign land, however, the stakes between interstellar communication are much higher than normal. There are several reasons for this.

The first is that both parties in an intercultural extraterrestrial communication have no real reason to pursue successful communication beyond their own curiosity or material advantage. A non-human cultural relationship will not resemble a friendly tourist trying to use hand gestures to order a beer. Each non-human communicative act, especially early in the relationship, will carry with it the weight of a real diplomatic statement, complete with perceived context and insinuation. A statement coming from a human to an alien race, in this case, speaks for the human race as a whole, whether we like it or not. This adds a degree of severity and gravitas to extraterrestrial communications; they will always carry with them significance beyond perception, and will leave little room for misunderstanding and impatience, beyond the limit of the negotiators themselves.

The second, and related, point is the natural disadvantage present in two interstellar species who view each other in a competitive manner with a questionable respect for the others' lives. This naturally causes errors to resemble insults, and insults then to incline to conflict. This admittedly pessimistic view of nature and inter-species communication has its roots in how two species, which are so completely different from each other, tend to see the other as an inferior life form. Whether this inferiority is perceived through culture, technological progress, or military capability, it does not matter. It only matters that cultures which are completely different have no frame of reference for comparison, which tends to lead one to believe their own culture is superior because the other is not understandable.

This built-in cultural obstacle, therefore, sets a limit on how far and how involved a diplomatic discussion can at first go. During human state-to-state interactions, diplomats can always fall back on the simple fact that both sides are people, and subject to the same needs, wants, desires, and laws of nature as the other, no matter how much they hate each other. Culturally, all humans are more alike than they are different in this regard. This notion can—and has—been used in the past to negotiate ceasefires and peace between human political groups.

The relationship between humanity and a non-human species will not necessarily be this way. A sad and sobering fact of extrasolar relations, as we may unfortunately discover, is there is no political incentive to respect a completely different form of biological life if that form of life is both completely foreign (in terms of species) and also a competitor for resources. Only our own subjective morality will be a reliable source of this respect, and therefore, of mercy. While history has shown a very clear pattern that true carnage begins when adversaries begin to view the other as something lower than human life, we have no reason to believe this unfortunate human trend will stop when we encounter extraterrestrial competitors. In this case, the interstellar culture in question will *in fact be* something other than human life right from the start—there is no need to "dehumanize" something which is already not human. Our species' best human butchers have shown the only real prerequisite to generating hate and sanctioning atrocities is to view a rival group as less than human. If interstellar communications begin with this fact *already* in place at the very outset, it is painfully obvious that, depending on the negotiators and their attitudes, there is a clear temptation to disregard difficult communication or to give up completely on frustrating negotiations already in progress. This, of course, is a clear risk to conflict. Compound this with the already-challenging aspect of communication with a species that does not necessarily favor humanity's preferred modes of communication (e.g., written documents or speaking), or is unintelligible based on it being non-animal life, and the inclination towards conflict and ineffective communication becomes clearer indeed.

The third, and final, main reason the stakes between interstellar communications are higher is the inherent distrust between parties which do not understand each other. Language, and therefore communication, is a tenuous thing—each species at its higher levels of sophistication is inclined to finely hone its particular chosen communicative tool, which adds to complexity and thereby increases the chances for misunderstanding. Words matter, and nuanced and symbolic language is regularly employed in international relations to deliver messages between states, and even today these messages are regularly misinterpreted or missed. When communicating with a non-human intelligence, this overdependence on one form of communication (nuanced messaging through deep understanding of individual words) could lead to distrust between species which cannot use or understand this form of communication. Consider the most famous speeches of Earth's most impressive orators. Each word, each sentence, is carefully crafted to produce an effect, to avoid misunderstanding, to leave opportunities open via vagueness and non-committal concepts, and to maneuver for the best possible advantage for the speaker's chosen cause. These factors, while significant to human political groups, will be useless upon first encountering an extraterrestrial communication method. All parties must go back to the drawing board of simple language, and against a background of mistrust and obscurity—what are they trying to say? Is it a trick? Can we believe them? If so, should we? Should we commit to future communication, or should we keep these people at arm's length? Is communicating even worth our time and effort?

To combat the onset of severe errors that could cause mission-ending or war-inducing interstellar incidents, we can benefit from terrestrial experience. U.S. Air Force pilot instructors, during advanced flight training, have for many years attempted to forewarn their students about the most common decision points during missions where errors are most likely to appear. The three main error-prone behaviors traditionally identified by flight instructors, in turn, can compound on the others and create spectacular failures in the air, sometimes leading to mission failure, loss of aircraft, or death. Worse, these error-prone behaviors could occur each and every flight since they could happen during the essential steps of a human's decision-making process. However, if these error-prone behaviors are understood prior to encountering them, pilots can approach potentially dangerous and fast-moving situations armed with the knowledge that their own mistakes could be their own worst enemy.

Errors in Perception

The first instance where errors can occur during decision-making, and during communication, is also the first opportunity chronologically. Above all, perception errors indicate a lack of or misinterpretation of information. Correct and useful information is usually very hard to come by during times of imminent or unknown threat, and would certainly be difficult to determine prior to discussions with non-human organisms. For our discussions in this book, obtaining correct and usable information to form a correct perception means good military intelligence. Beyond simple communication, sizing up an extraterrestrial threat will likely err towards the side of overestimation, not an uncommon thing during intelligence estimations. We have an interesting example here from Earth history: the Cold War.

During the Cold War, U.S. and Soviet intelligence professionals were constantly estimating the others' capabilities for obvious reasons. This led not only to complacency, but interestingly it led to both sides confirming within each other's camps that their opponent was somehow more dangerous than they actually were. Publications from western media and think tanks at the time display their unvarnished pessimism regarding the military power of the Soviet Union and Warsaw Pact nations aligned against NATO. Even in the glory days of the "sovietologists," the professional linguists, academics, and so-called cultural experts employed in great numbers in a great many government offices to peddle their knowledge and familiarity of Russia, the number of times the experts got their Soviet perceptions wrong is noteworthy.

Consider one famous article by John Mearsheimer written in 1982 and published in *International Security*.[1] In it, Mearsheimer explains to the reader why the Soviets and accompanying Warsaw Pact nations would have a harder time invading Western Europe than commonly thought. In fact, his very thesis seems to exhort the public to understand a Soviet attack *would not actually be that easy*.[2] This seems obvious today; of course rolling through 500,000 allied troops equipped with anti-aircraft weapons, armored vehicles, and a massive air advantage would not be easy. But the incorrigible prevailing "woe are we" attitude in the west at the time necessarily lent itself to pessimistic (and often unfounded) views that the Soviet Bear and his Central European slaves were ready and capable of thrashing NATO forces.

Today, such estimations of infinite Soviet strength and unlimited cunning look error-prone indeed. However, for those who lived through it, it is easy to understand the prevailing feeling at the time. At any moment, if the Soviets (or the United Sates, for that matter) felt a little gutsy, or maybe even got a little too drunk, World War III could begin at the drop of a hat. A twenty-minute missile flight time is all that separated life as we knew it from charred remains—and still does today. This constant condition creates a "gun-to-the-head" syndrome. While it may seem abnormal to accept imminent Armageddon as a day-to-day occurrence, it is not; and interestingly, behavioral defects like this are more common to human life than one might first think. A large part of how humans make errors in perception comes from these defects.

One of our more prominent brain defects, likely evolved as a survival mechanism, is called "cognitive bias." This particular flaw in our gray matter causes us, among other things, to assume the worst when we see an otherwise explainable sign of trouble. One of the most classic examples, as related by Michael Shermer of *Skeptic* magazine, involves our most basic instincts.[3] Consider a pre-historical hominid travelling alone in the African savanna, relates Sherman. Suddenly, the bushes nearby begin to rustle. The hominid now has two choices: assume the rustle is a predator, or assume the rustle is just the wind. If the hominid believes the rustling to be the wind, and therefore no threat, he is essentially taking a risk that he is right at the potential cost of his life. Consider the opposite reaction: the person instead fears the worst and assumes the rustle is a dangerous predator, even though it may not be. Shermer calls this a "Type I cognitive error," or false positive; even though the hominid misperceived the rustle for a predator, no harm was done, and he went about his day very much still alive. The contrary example, though, could cost someone their life, and therefore would prevent them from passing along their genes to any offspring they would have had. Shermer calls this latter error a "Type II cognitive error," or false negative—asking yourself "what could possibly go wrong?" before falling off a cliff.

Why does this matter? As it turns out, humans get worse at assessing the difference between committing a Type I or Type II error the more high-intensity a situation becomes. This means, as Shermer also relates, our human pattern-detection skills, which are usually reliable when no threats occur, suddenly become unreliable in life or death situations. This leads people to take the safest course: that all rustles in the grass must be predators. After all, if it turns out to be nothing, one can simply laugh at oneself and stroll away. You can see where this could become a problem in an interstellar theater; a galaxy-sized "rustle in the grass" could generate intense feelings of panic and war preparation that may be nothing more than the solar wind.

Errors in Judgment

Even after committing an error in perception, all is not yet lost. One still must act on these errors to see their black blossoms come to fruit. Before things really get bad, the point at which errors occur during analysis is called an error in judgment. We all have committed these in our everyday life. For example, we have a bad experience with a coworker and decide with certainty they must be an intolerable ass who hates us; until we find out they have had a family tragedy, or they simply had a bad day during your first impression, or they happen to be mentally ill. Where seconds before we were sure they were a hated foe, we suddenly come to realize they are a victim of ill fortune. Errors in judgment are the source of many of our cognitive errors.

The fact is, we never have all the data we ever need, and in many cases, we often have too much useless data to reliably sift through in order to make a solid decision. When there are too much data and too many distractions to quickly make a reliable decision, errors in judgment happen because time constraints prevent a thorough analysis. Given infinite time, it is much more likely any decision we make will succeed, or at least come as close as we can get. However, major decisions, notably during time-compressed military operations, can almost never be given their due diligence. This does not always mean a high-intensity combat situation. Even everyday tasks which would challenge any leader find themselves troubling, and compounded with a great variety of different tasks, most leaders usually do not have enough time to thoroughly analyze every decision they make. For example, a military unit needs to upgrade a critical piece of equipment, and won't likely get another chance to upgrade in several years; and the selection must be made without enough information about its long-term effects. In another example, when ordered to reduce personnel in a particular military unit, how does a leader make the selection? Does he pick the troop who is most damaging to morale, or does he pick the least effective one? How will this affect the big picture in the next three months? Six months? Two years?

These questions are not easy, and when leaders are forced to make judgments under the tyranny of the clock, or with poor understanding, bad outcomes will naturally occur.

Errors in Execution

The last chance to halt an error is just prior to and during execution. At this point in a decision chain, the decider has either satisfied his own information requirements sufficiently enough that he believes he can make the right call, or he has run out of time and must execute a course of action. In the latter case, there still exists the possibility of indecision, which is almost always the worst choice and usually results in catastrophe. In the former case, there is almost nothing which will stop the decision maker from executing his chosen decision, save a last-minute powerful form of evidence. Put another way, an error in execution is simply making the wrong choice.

These three steps apply to dealing with non-human entities because each situation will no doubt carry with it insufficient information, will be pressured by time, and will involve decision makers prone to cognitive biases. A fictional example is the best way to show how the three-stage error process can make things go wrong.

Imagine a fighter pilot is patrolling a no-fly zone over a country which was recently an enemy. Tensions are high following the recent ceasefire, and the former enemy resents the no-fly zone as a violation of its sovereignty. It is well known the enemy will violate it every chance it gets, if not to assert its rights then to simply prove its point that it is not truly defeated.

One day, the pilot picks up the radar signature of another aircraft, which is displayed as "unknown" on his scope. Regardless of the data which is available to him, the pilot is the ultimate authority when deciding what to do about this contact. To make this decision, first the pilot accesses his known information—the limits of his knowledge which sketch out the situation as he understands it. After accessing his current affairs knowledge and pre-mission study, the pilot knows he is currently flying a combat air patrol in a no-fly zone. Further, he also knows the same Identify Friend or Foe (IFF) codes he has equipped on his aircraft have been passed out to all friendly aircraft; this means every aircraft in the no-fly zone should be displaying these codes. After checking his information, it appears the unknown radar contact is not displaying these codes. He then confirms there are no known IFF equipment failures reported by other friendly aircraft in the area which could explain the "unknown" status. He also understands aircraft are not authorized to takeoff without this code, lest they be blown out of the sky under the hard-and-fast rules which govern a no-fly zone.

After this first general step, the pilot reaches for additional information: situational awareness. He looks at the speed and direction of the unknown contact. Is it in a hurry? Where is it going? What is its altitude? It happens to be daytime; why would a lone enemy fighter be out here now? The unknown contact's attributes in this regard can reveal clues about its intentions. The pilot checks again: the unknown aircraft is flying quite slow, but not so slow that it couldn't be an enemy fighter. It is also coming from the general direction of a former enemy airbase. His heart begins to beat faster. Could it really be a no-fly-zone violator? Are they truly that stupid to sortie against us in broad daylight, and alone? And then, finally, his ego begins to enter the equation: could this finally be the day I do what I've been trained to do?

The fighter and the unknown aircraft continue to close. The fighter pilot asks his wingman if he sees the same thing he sees. The wingman says no. He confirms with airborne command and control if anyone ought to be out here. The command and control replies that they see nothing. No more information is forthcoming; the pilot must decide on how he will perceive this potential threat. If the threat is an enemy fighter, he has to accelerate now to have enough speed to enter an air-to-air engagement with enough energy to given him a lethal advantage if they begin to try to outmaneuver each other. Faster speed makes the closure that much quicker.

The pilot orders his wingman to "push it up" (increase speed) and prepare for a fight. He informs every conceivable airborne command and control asset of his decision, who reply they acknowledge his decision but reiterate they do not have the enemy contact on radar yet. The pilot checks his fuel level and glances at the weather one last time. His fuel will allow for about a 15-minute engagement, more than enough time to slay his foe. There are some thunderstorms moving into the area, but they still look small. Conditions are looking good for a fight. The wingman is keyed up—he's only been in country for five days and already thinks he's going into his first air-to-air engagement in a world where they simply do not happen much anymore. The proverbial rustle in the bushes rustles louder.

The fighter pilot continues to crank through his calculations. He chooses his weapon and plays through the engagement in his head. All the while, his brain's logical processes try to determine what the odds could be that this enemy is real, out here in the face of idiotic odds. Why would it be out here in daylight? Its IFF must be off, or malfunctioning. Why would someone fly through a no-fly zone with no IFF switched on? Why can't airborne command and control see it on radar? This guy is all alone, too; would the enemy really be stupid enough to sortie against us in a solo aircraft when they know we always fly in pairs? None of it makes sense. But what if we're wrong? Then the enemy gets the jump on us—then we die. What if the bushes really *do* contain a tiger?

The pilot convinces himself he's accurately determined the unknown contact to be an enemy. All aircraft continue to close at breakneck speed. Something is still nagging at him; must be pre-combat jitters, he thinks. But he knows what it really is: the lack of data about what this aircraft could be makes him uneasy about pulling triggers without making absolutely sure. There's only one method left to satisfy this nagging doubt: a visual reconnaissance of the aircraft. Naturally, if it turns out to be an enemy aircraft this is an extremely dangerous decision. He tells his wingman what he wants to do. The wingman responds affirmatively but hesitantly. He knows his life is on the line too.

The fighter pilot can now make out the aircraft in the distance. It's at about 28,000 feet, 400 knots. A bit low and slow for an enemy fighter, but still a definite possibility. Just then the airborne command and control aircraft radar operator, now excited by the impending action, interjects on the radio to tell the pilot they are now tracking the unknown target on radar. Hoping for more data, the fighter pilot asks for more information, but the command and control aircraft responds they know nothing more. The command aircraft goes so far to ask *him* for more data. The pilot swears under his breath and silently thanks them for the free confusion. He finds himself distracted from the business at hand. The pressure is building; the intensity makes it difficult to think clearly.

The enemy aircraft comes within infrared missile range. It was earlier in radar missile range, but the decision to obtain visual recognition convinced the fighter pilot to refrain from giving away his position by targeting the enemy aircraft with a radar lock. Were it truly an enemy, it would surely pick up the lock and reveal his intentions. Both aircraft in the friendly formation prepare their infrared guided munitions, designed for short to medium range engagements. The aircraft comes into view.

"Stand down," barks the formation leader over the radio. As soon as he saw the aircraft's silhouette, he knew today wouldn't be his day in the sun. The opaque and formless outline of the enemy aircraft gave way to a clear design revealing it to be a business jet. As the fighter pilot zooms past the interloper, he notes the now too-familiar red and gold flag of the People's Republic of China. Looks like it was yet another Chinese businessman fleeing the war zone in his private jet. As it turns out, it was a poor decision by the Chinese government's state-run industrial complex to heavily invest in the war-torn country, especially after being counseled by Western countries to cease their industrial exploitation activities. As usual, the Chinese government had neglected to inform the western coalition of their movements and decisions, underscoring their claims that they do not recognize the coalition's no-fly zone. Third time this month. The fighter pilot conducts an intercept, and after a sternly worded message over universal frequency, the Chinese aircraft changes course. He'll never know how lucky he was. The rustle was just the wind after all.

These kinds of situations can and do happen. Here you can see both errors in judgment and perception, but an error in execution avoided at the last minute by the prudent—but risky—decision to obtain visual recognition of the "enemy" aircraft. The lessons here are very clear, and certainly apply to future miscommunications with non-human entities. The perceptions and emotions of each party helped guide the situation towards disaster. The fighter pilot, inculcated from his first days in training to seek and aggressively destroy enemy threats in the air, not only sought combat but *wanted* the enemy aircraft to truly be an enemy. His wingman was no less effusive. The Chinese aircraft operated under a policy which scorned the real combat situation for a preferred but illusory political reality, putting their lives at risk and potentially pitting their nations against each other in war. The command and control aircraft's last-minute radio call injected useless data and confusion into a rapidly developing situation. Most importantly, the *further* the situation progressed, the *harder* it was for either party to pull back from the brink. It was only the fighter pilot's experiences and gut which told him something could be wrong. Note that it did not tell him something *was* wrong; indeed, in another scenario with a different pilot, that sense very well could not have been there to help.

And so it will be with non-human relations. The above scenario's critical components are limited neither to species nor to thought process. Indeed, the fact that the above scenario dealt with all members of the same species actually worked to promote a bloodless solution; we may not be so lucky if the next business jet happens to be a spacecraft flown by a non-human. In this example, both parties understood the other's basic standpoint, the tactical situation and associated risks, likely reactions, and the possible outcomes. When encountering a completely different culture, though, this will not necessarily be the case. The key will be to approach each situation with caution and deliberate slowness. The more time there is between closure, the better the chances of understanding and survival.

Superior Civilizations

Dealing with a superior spacefaring culture is rather simple: deal with them as little and as far away as possible. By "superior," here we mean to say a culture—a civilization—which has reached clear technological and/or intellectual superiority over ours, resulting in a complete imbalance in power. This would be a very bad thing for the inferior party.

What makes another spacefaring civilization superior? It is not as easy to articulate as it first sounds. There are obvious signs that a civilization is more advanced than ours—high technology (especially flashy technologies like weaponry, energy production, and advanced spaceflight propulsion methods), a larger population, even the trappings of some kind of spacefaring empire. But the threat may be more nuanced than expected. For instance, a newly discovered civilization can be more politically advanced, and have a politically unified homeworld, which happens to be an especially dangerous signal that superior military power is just around the corner. They could possess a higher degree of cultural sophistication, or a non-military technological superiority which could threaten our interests and culture.

As long as a civilization remains superior to ours, we know they will always be a physical threat to our survival. This will drive many hard discussions in the statehouses of our primitive nation-states here on Earth. Will this culture attack us? For what purpose would that serve them? How far advanced are they compared to us? Would they be capable of launching a planetary invasion, and if so, would resistance be appropriate? What would we as a species be willing to surrender to save our species in the event of overwhelming force? If they are friendly out of some benign sense of curiosity, as a child views a frog, what does that really mean for us? Could we obtain their technology and catch up? What would they value in trade if we wanted to engage in commerce?

Clearly, the best strategy in the event of encountering a superior civilization is to focus on diplomatic interplay vice military confrontation. Any peaceful strategy which lowers the chances of war would be worth our time. In this instance, finding methods to communicate and sharing cultural information as soon as possible should be our top priority. This is primarily to forestall armed conflict by communicating to our non-human superiors that we are worth preserving, and perhaps could even offer assistance in their own interests. If sharing our society and culture leads to some kind of friendly political arrangement, so much the better. It is likely the superior culture would find something useful to them here on Earth, or better yet, in our solar system (so we do not have to surrender terrestrial resources). An exchange of personnel and cultural resources has not only the benefits of allowing us to surveil them, but is basically free to both parties with no commensurate military commitment. Above all, in times of discord we should play for time and stall when possible, offering our choicest diplomatic excuses and promises. Presumably, distances between our territory and theirs would be quite distant, which adds buffer space and drives difficult decisions into the rival civilization's decision matrix should they begin thinking about invasion or conquest.

Inferior Civilizations

A rival's inferiority grants options to the superior. Paradoxically, dealing with an inferior culture will be more complicated and difficult than a superior one. Why? In the latter, our only intent will be survival; in the former, we must decide what our intent will be. And in our species, that requires consensus.

First and foremost, we will have to make the determination that the inferior civilization meets our definition of spacefaring at all. Once this is done, we will need to decide if we wish to make contact or not. While it is not always true a superior civilization will find the inferior one first, it is likely that if our civilization is more technologically advanced than the inferior one, we can probably detect their presence before they detect ours, and in greater detail.

Perhaps the most famous inferior civilization contact policy of all is *Star Trek*'s infamous "Prime Directive."[4] Featured in numerous episodes of the famous television series, this policy forces any vessel or personnel to refrain from contacting an alien species if they have not yet attained technology capable of propelling their vessels at faster than light speeds. Impressively, by including this thought process the show's creators displayed remarkable foresight at what could go wrong not only when conducting a first contact, but that contact's subsequent fallout and consequences.

While there is wisdom in the policy, there are also ethical and security considerations. As a superior species, refraining from interceding on an inferior's behalf could doom them to potential political or environmental enslavement or destruction the superior civilization could theoretically prevent. If the superior intervenes, its intercession would most likely act as an unpredictable stimulus into a culture unready to face the reality other sentient cultures exist. This is not a difficult situation to foresee. Imagine, for example, in the future an advanced humanity encounters a species which is roughly in the same technological state that we are in now: a gradual decrease in endemic warfare with slowly improving global economic ties, and a society which is suffering from the clear environmental consequences of planetary-wide industrialization. If we observe this species to have degenerated into nation-state political squabbling which completely blinds them to the ominous environmental impacts of industrialization, we would have a clear choice: inform them of what could happen if they do not repair the damage to their environment, or do not.

If we do, by revealing our presence we would force a chaotic political change onto the civilization—not many political systems can simultaneously cope with the knowledge they are not alone in the universe, that those who proved this are technologically and militarily superior, and that these aliens also provided conclusive evidence that their planet is dying. The results would be unpredictable and dangerous. Would the local culture see us as benefactors on high ready to cooperate with them, or as hostile aliens who are well ahead technologically and are an obvious existential threat? Would they bother with the environmental concerns we informed them about in the first place if they are suddenly confronted with news that shakes all political systems to their core? Exactly how many governments would remain standing under the impact of an intelligent alien visitation to the world? And, by the way, which government would we decide to contact first? Does the planet in question possess a pseudo-United Nations? Should we contact the strongest government for the greatest impact, or the weakest one to guarantee the least blowback? How would our perceived favoritism play out on their global political system? Would the planet somehow pull itself up by their bootstraps out of fear in the face of our technological superiority, powering a technological catchup which pits them against us as an adversary forever? And would our salutary observation allow this catchup without falling into a Thucydides Trap?

These are just some of the nearly limitless permutations facing a first contact situation. It was these questions and their complicated possibilities which drove the *Star Trek* creators to develop the Prime Directive policy. Indeed, the injunction which prevents contacting other spacefaring civilizations until they are capable of standing on their own two feet has excellent merit for a superior species which is legitimately interested in peaceful discovery and contact with other cultures. If conquest is our aim, however, there is almost never a good reason to wait, and a policy resembling the Prime Directive would be useless. In summary, an inferior civilization must be watched carefully until the decision to make contact is finalized.

Inter-Species Intelligence Gathering Limitations

How does one spy on an adversarial species which is not one's own? The question raises critical intelligence challenges that have never applied to terrestrial problems. Unlike terrestrial adversaries, there will be no free mixing between species once we encounter our first non-human sentient being. There will be no common culture, nor probably even a common atmosphere to breathe. There will be no embassies to staff; no parties to host; no trade to conduct at first. In short, observing a species from afar will more resemble an experiment in a beaker than a true international relations problem.

On Earth, states gather intelligence on their adversaries in several ways. Traditional spying, of course, involves surveilling a person or organization without their knowledge, and usually within their own territory. Military reconnaissance, on the other hand, is not considered espionage but is called surveillance; the difference being military forces watch each other from outside the target's sovereign territory. Aside from these methods, the differences between terrestrial and extraterrestrial intelligence gathering essentially lie only in the tools used: humans, satellites, spacecraft, listening posts, and so on. When it comes to interstellar intelligence gathering, we should expect the complete absence of what is now called "human intelligence"; those factors which can be learned by individuals, usually agents, after placing themselves among the adversary's population and physically interacting with them.

The dearth of human intelligence available to either side of a looming interstellar conflict is stark. There will be no established embassies; no individual trust of any kind; no mutual cultural understanding; likely no ability for either species to comfortably exist in the other's environment for long periods of time; and no sexual trickery, which we find so effective here on Earth. In short, a different species will be so unlike us, the very nature of interacting with them will defy all previous human intelligence-gathering techniques.

However, these are no reasons not to try. Intelligence will, as it always has, be absolutely essential to managing potential conflict. Indeed, intelligence must be gathered even before first contact is made. In the end, intelligence is the first line of defense when dealing with a potentially hostile enemy elsewhere in the cosmos. Below are some of the most pertinent considerations which will challenge our thinking about intelligence against a non-human potential foe.

Fleet Composition and Strength

You might call this data the main concern of any military or strategic planner. The most important factor when considering potential strength is pure military power; and that power is held in the adversary's spacefaring fleet. This concern includes classic hard power measurements: number and type of vessels, personnel (living or automated), technical makeup, preferred doctrine, and weaponry. Any scrap of information can lead to conclusions about the adversary's fleet specialization strategy, which in turn will reveal hints about how they may deploy for battle. Traditionally, spies and espionage are best employed for this kind of information gathering, but against a non-human threat we will need to think creatively.

Communications Espionage

Since it is not easy to simply walk onto an adversary's planet and blend into a crowded alien marketplace, grabbing as much intelligence from a distance is key. For this problem we turn to the age-old art of signals intelligence. "Signals" in this case go beyond traditional flags and pennants used by terrestrial armies and navies. Rather, it refers to communication conducted in the electromagnetic spectrum, and includes everything from cellular phones to television, from radar arrays to electromagnetic signals given off by weapon systems as they are powered on.

Collecting information on a spacefaring adversary's communication systems is best done with passive systems to prevent any chance of detection. First, electromagnetic waves leaving their territory should be captured and analyzed for comparisons to our equipment to best determine the waves' most likely source and purpose. Second, any civilization that could be considered a threat to ours will almost certainly boast a robust satellite system in orbit around their homeworld. This system should be electronically penetrated (and hacked, if possible), and as much information as possible should be collected. Decryption, if necessary, can be done later. It may be helpful, but risky, to affix probes to these satellites or establish nearby listening posts, out of sight from the adversary's prying eyes.

If successful, communications espionage stands to give us not only the first images and basic information about a potential adversary, but also critical cultural, scientific, and operational knowledge about the inner workings of the species. Surveillance in this way should be as lengthy as possible; it is doubtful cultural trends and language can be deciphered without a surveillance period of several years or more.

A Note on Intelligence Fiefdoms

A sad reality of the intelligence community writ-large, in any nation, is the tendency to inadvertently separate the various "types" of intelligence into self-governing and separate communities, like separate compartments in the same vessel. This paragraph is a word of caution against this practice. Space is vast, confusing, and complex, and the limited professionals we have now will never be enough to satisfy a real operational problem should the need arise. Deep space operations require a decision-making process made more vulnerable by the vast resources and time needed to execute, which means intelligence assets and information must be unified as much as possible. Non-military readers will likely be surprised at this notion; after all, popular film make intelligence gathering look shiny, professional, and unsung. Unfortunately, in truth the unity of any intelligence enterprise currently depends largely on the preferences and traditions of that particular organization. This will not be good enough for space operations. In space, there is no human intelligence, no signals intelligence, no electromagnetic intelligence—there is only "intelligence." And the stakes will never be higher to cooperate, or potentially lose everything.

This chapter discussed key considerations with dealing with non-human competitors, should we ever encounter them. While the odds of this occurring in the near future remain very low, and the discussion remains largely academic, it is still instructive to examine the possibilities associated with an encounter with a non-human sentient civilization. From this discussion, we can also see several weaknesses of human cognitive decision-making, as well as several techniques which could aid any fight against any spacefaring enemy. By examining what could happen against a non-human foe, the academic exercise leads us to several conclusions about how to think about intelligence and decision-making during deep space warfare.

8

Likely Causes of Warfare

> "You know as well as we do that right, as the world goes, is only in question between equals in power, while the strong do what they can and the weak suffer what they must."
> —Thucydides, *The Peloponnesian War*

Given the costs, risks, and national effort involved in deep space warfare, the question of motivation is quick to surface. Why would a nation or culture embark on such an expensive and potentially risky ordeal, and what would they hope to gain? We already know the answers—interstellar warfare will be conducted for the exact same reasons we conduct warfare today.

Resource Competition

One of these reasons, and the most oft-cited cause of alien assaults on Earth in popular science fiction, is to seize material resources. However, from a strategic standpoint resource competition is actually one of the *least* likely, and one of the worst possible, reasons for going to war.

There are several reasons for this. The first and most obvious is the sheer amount of unclaimed and freely-available resources in the cosmos which require no combat or struggle of any kind to obtain. As stated above, NASA estimates the value of the materials found in the asteroids in our solar system's asteroid belt alone to be a gargantuan $700 quintillion.[1] That is no exaggeration, nor is it a fake number; in fact, such material wealth is enough to distribute $100 billion to each person now living on Earth. Even if we found an inferior extraterrestrial culture and had the means to travel to and conquer them, there would be no reason to do so for resource collection purposes with the asteroid belt in our own backyard, not to mention other resource-rich areas in unclaimed and uncontested stellar systems.

We here on Earth like to think we are special; that there is something about our amber waves of grain, blue oceans, and nitrogen-oxygen atmosphere that would entice an alien assault force to claim it for their own. But the reality is much more humbling—Earth is simply constructed from the same materials that populate the cosmos, and its chemical composition reflects this fact.[2] There is no reason to believe an alien culture would have to assault Earth to obtain anything unique at all, unless it was planning on filling an interstellar zoo with samples of Earth flora and fauna. Would an alien strike force really travel the tens to thousands of light-years necessary to begin a risky military campaign to claim our comparatively low amounts of strategic earth metals, most of which are already being consumed by humanity? If they needed something plentifully found on Earth, like water, would they really forgo the relatively clean and readily available fresh water-ice found in untold numbers of comets and planetoids, only to attempt to steal the comparatively dirty and salty oceans of Earth? If they needed oxygen, synthesizing it from readily-available elements found throughout the cosmos—and closer to home—would be many times cheaper than launching an invasion fleet.

In other words, why would an invasion force—human or otherwise—travel extraordinarily far to violently wrest a questionable supply of limited resources from an unpredictable enemy who has likely consumed most of these resources anyway? Why do this when there are much easier, closer, and cleaner resources available with no realistic chance of foreign interference?

It is conceivable that in some point in the distant future, Earth could find itself in competition for a particular space-borne strategic resource with another spacefaring culture. This could be a particular element necessary for energy production, a gas only found in a particular nebula, or other such rare element which really is found in low quantities in the cosmos. However, in this case the size and scale at which these astronomical phenomena are found, like nebulae, black holes, and asteroids, immediately poses a counter-argument as to why two species would have to fight over a single resource. The closest nebula to Earth, the Helix Nebula, for example, is approximately 2.87 light years in diameter and growing at the rate of approximately 31 kilometers per second.[3] It contains a treasure-trove of Helium and Hydrogen, as well as many other gases given off by the death of its former host, a white dwarf star, about 10,000 years ago. Given the large amount of material in the nebula, it is not only reasonable that several civilizations could mine this nebula at the same time with little to no interference with each other, but it is also unreasonable for one culture to expect to be able stop another from doing so due to the sheer size and scope of the expanse.

Territorial Disagreements

If resource competition is not a viable reason to go to war, territory might be. Just like here on Earth, political rancor concerning territory will be the most likely tinder which lights the fires of war. The sources of these political disagreements are well known to humanity, and as mentioned above it may be necessary to strategically analyze space through the lens of our historical experience here on Earth. In the case of a non-human adversary, there is only the matter of adapting our terrestrial thinking towards their culture, which in turn is only a matter of intelligence, study, and time.

We should expect territorial disagreements and encroachment to be an issue in future spaceborne political woes. The first step in facing this fact is understanding we possess a primate-driven concept of territorial integrity which now and will forever color our political perception of territory. Every primate species on Earth concerns itself with property and territory; the more violent the species, the more they are willing to fight for and seize foreign territory. Primate concepts of possession, which we can see often in concepts of slavery, "rights" of the conqueror over the vanquished, and religious traditions which shape our conceptions of property, like the Hebrew Bible's Ten Commandments,[4] spill over into daily politics and will certainly do so in the future.

While it is not yet clear how our species will view territory in space, it will most likely be thought of as ownership over star systems and access rights to the places and spaces between different systems. As we noted above, star systems and stellar phenomena, belonging to the approximately 5 percent of space which is not military emptiness, represent the parts of deep space in which humanity will have vital security interests. Logically, this means these are the portions of space we will likely try to claim. Along these lines, access to these locations will also have to be guaranteed as part and parcel of system ownership.

This brings us to an extraterrestrial culture's view of territorial integrity. Given the nearly limitless cultural combinations possible, there is a chance that any spacefaring civilization we encounter will prioritize territorial issues completely differently than our species does. However, it is much more likely we will agree on the value of territory and its importance for a variety of reasons. First, any other spacefaring civilization would probably need to have evolved in a highly competitive environment on its homeworld, similar to humans. The very fact that a civilization is spacefaring means the nations or cultures have successfully advanced to the point that they are *capable* of space flight. This also means said civilizations can comprehend and develop complicated socio-political and diplomatic institutions, and they have likely defended these institutions and organizations from their enemies in the not-so distant past. Species which reach the level of technological sophistication which includes spaceflight do not get there by accident; only intense competition for power and resources, along with endemic struggle and warfare, are the only likely forces which can propel a culture to the dizzying heights needed to escape its planetary bonds. It would be foolish indeed to assume a non-human spacefaring civilization would not understand or possess concepts of sovereignty, and that these concepts would not be attributed to deep space territory.

Fear

As our previous Athens-Sparta example from *The Peloponnesian War* revealed, the fear of a rising rival or of an enemy's power is always enough to provoke war. Fear brings with it irrational thought and panic, and fear's primary offspring is doubt about what *could* happen, not about what *will* happen.

Wars begun out of fear begin with one question: "*what if we do nothing?*" In other words, wars which begin out of fear often have aims designed to forestall a rival's rise or prevent some kind of threatening foreign military or political ascendency. They are therefore often undertaken at the last possible political minute, or are otherwise too late, and usually after other political options have been exhausted. This means conflict caused by fear can often be seen coming before the decision has been made.

Besides the Peloponnesian War, one famous example includes the Mesoamerican native tribes who had been vassalized by the Aztecs in pre–Spanish Mexico. Once these tribes realized they could slough off their master's yokes by co-opting the Spanish *conquistadores,* they went to war out of fear of their Aztec masters who regularly exacted tribute and took their people as sacrifices and concubines. They calculated any power would be better than the Aztecs, which in the end turned out tragically; but it does indicate the power fear has in making life-and-death sovereignty decisions.

Honor

Honor is always relative to the culture and species. As far as we know, honor is a uniquely human cultural aspect which comes from our primate-inspired valuation of territory, possessions, and social needs. There are plenty of creatures on Earth who have no need for honor. For example, ants and bees, being hive-driven "societies," simply perform their own particular function with neither offense nor apology, and care nothing of credit. If one of their number falls or fails, another takes its place with no social repercussions of any kind. Because humans depend on social groups to survive, how these groups view individual humans have a direct effect on whether that particular human and his or her offspring live or die. For instance, one prominent psychological theory is that rice-growing cultures are more communal in nature because of the need of the entire community to cooperate in order to care for the wide-ranging needs of the plant during its growth cycle. Psychologist Thomas Talhelm conducted a research study in 2014 where he compared the personalities of wheat-growing northern Chinese with rice-growing southern Chinese, with interesting results.[5] According to the research, rice-growing cultures tend to depend more on their local social groups and their community than strangers, and tended to view themselves more often as a small piece of a larger machine than as an individual.

These cultural differences matter. They help explain the almost pathological need for respect, admiration, and status which humans crave, along with our quickness to take offense and seek retribution. Each step up the social ladder brings greater survival through better security, but also brings intense competition to secure this new status and safeguard one's reputation to prevent it from eroding. It is no wonder our ancestors sought duels to satisfy their social honor; to lose it would mean a slow death by ignominy, or at the very least life as they knew it would be over.

One famous example of a war heavily influenced by honor was World War I. After the assassination of Austrian Archduke Franz Ferdinand by a Serbian radical, Germany was forced to back its treaty ally Austria-Hungary in demanding satisfaction from Serbia, even though the perpetrator was not tied to the Serbian government. When Serbia acceded to nearly every Austrian demand, some of them humiliating, in order to avoid war, Austria-Hungary attacked Serbia anyway, thinking it would be an easy victory. In this case, Germany's honor compelled it to stand by its military ally rather than extricate itself from a political situation where Serbia was obviously willing to satisfy its neighbors in lieu of war.[6] Germany's decision to stick with its ally despite dubious reasons for going to war was in no small part due to an incident a few years earlier in 1911. Following threats of war over a disagreement between France about Moroccan territory, Germany felt its honor was besmirched when it was compelled via international pressure to back down from its position.[7] Three years later, the resulting four-year horrific conflagration clearly shows honor matters to humans.

Self-Interest

This category includes wars begun for conquest and national gain at the expense of another state. While many wars are fought for self-interested reasons, self-interest in this case means the zero-sum transference of something from the conquered to the aggressor. Since before recorded history, humans have made war on their enemies to take their produce, possessions, and people. In many ways, war was first an exercise in efficiency; if you can take for free what your neighbor labors to produce, that means better security for you and your neighbor is therefore less prepared to fight.

Projected onto a deep space environment, systems held by a rival would be the likeliest objects of conquest in any interstellar war. A close second would be stellar phenomena which possess excellent scientific research potential and chemical resources, and are less territorially distinct.

Examples of wars fought in self-interest are too numerous to count, but include the campaigns of Xerxes the Great, Alexander the Great, the Punic Wars, World War II, and the Korean War, to name a few.

This chapter introduced several reasons why humans fight, and the applicability of these concepts to future wars in deep space. The three reasons listed here are simply the classical representations of political realism, a political philosophy which is widely accepted but by no means the only political philosophy in existence. While the ultimate reasons humanity will go to war in space will probably be the same for why we go to war here on Earth, like all wars those in deep space will without a doubt compel political organizations to take great risk before seeing military success.

9

Challenges to Diplomacy

> Politics are not a science based on logic; they are the capacity of always choosing at each instant, in constantly changing situations, the least harmful, the most useful.
>
> —Otto von Bismarck

Diplomacy will always remain an integral part of warfare. It is a common myth that diplomacy halts when the guns begin firing; in reality, diplomacy always continues, even if only through back-channel and clandestine methods. Only through diplomacy can political entities truly negotiate with each other, capitalize on military activities, and make policies which decide the fate of conflicts. Indeed, the only way to end a war is when political authorities agree to end it, which can only be done via diplomacy. This requires agreement, and without sufficient military force to convince an enemy to succumb to one's political will, the war will not end as expected. As Niccolo Machiavelli so aptly wrote, wars begin when you will, but do not end when you please.[1]

Diplomacy with non-human cultures brings with it unique challenges. While some of these will be discussed here, there can be no doubt that unexpected and complicated problems will go hand-in-hand with deep space diplomacy, whether or not the adversary is human.

Biological Impediments to Interspecies Communication

There is no guarantee we will be able to communicate with a non-human culture. This is primarily due to the baffling variety of communication methods potentially available to living organisms, and due to the long, frustrating, time-consuming process needed to decipher each other's language. Unlike science fiction, there will likely never be such thing as a "universal translator"; the sheer variety, patterns, and combinations of speech makes this unlikely.

Moreover, other forms of biological, physical, and chemical communication methods available to living beings make this even more of a challenge.

For once, inter-species communication is a topic which has been approvingly handled by science fiction, probably because communication difficulties are a universal problem here on Earth and impact nearly everyone's daily life. Even when working in the same language and culture, communication is a task that requires constant attention, possesses multiple dimensions, and is an academic discipline all its own.

There are numerous examples of biological impediments to communication between our species and a non-human one. For one, not every species uses spoken language as communication. For another, while we can be reasonably certain any spacefaring species we encounter will be capable of communicating and recording thoughts and ideas via some method other than memory, this method does not necessarily have to be the written word.[2] The environment in which a non-human species evolved will essentially dictate the form of communication it prefers. An aquatic species, for example, will likely have evolved a form of sonar or sound communication which simply will not work in atmospheric environments. Other cultures could boast chemical communication, requiring a compatible physical interface between itself and its conversation partner. This of course may not be organically possible with humans. Then there is the matter of inorganic life, if we ever discover such a thing; how would they communicate? Indeed, a species which communicates in a similar manner as our own would be a great and lucky advantage if it turns out to be true.

Searching for Commonality and Technological Augmentation

Even if communication is possible, there must be something commonly understood between species to begin the conversation. The trick to finding common ground on which to communicate, regardless of method, is to focus on concepts common to all living things, then expand on language patterns once they become clear. A language, after all, is simply a logically structured pattern to express ideas and communicate thoughts; we can therefore be certain that whatever method an extraterrestrial culture uses to communicate will possess structure, grammar, and distinct patterns which express concepts and ideas. Once this is understood, communication will be soon to follow, even if it requires technological augmentation to make communication a reality.

It is this last point that makes interspecies communication potentially so difficult. In moments of heightened tension requiring fast decision-making, or at times where confusion reigns and messages become garbled or misunderstood, intricate technological solutions to communication problems may not be enough for the task at hand. Delays and doubt between a species whose preferred mode of communication is drastically different than ours could create major diplomatic concerns.

Above all, communication must be attempted, even if it consistently fails. Without communication capability, two different species are essentially doomed to an eternal Cold War; silently staring at each other from across the void, each is free to develop irrational and conspiratorial opinions about what the other may be up to. Unable to argue with each other except through force—the universal argument—will tend to compel diplomats on each side to begin looking at their relationship through the lens of force. Even if force is the only available way to communicate, an argument can be made that it would still be better than nothing, as long as it remains mostly for show. In this instance, the relationship would be like two silent players across a chessboard, making their decisions known to the other not through speech but through movement of forces, continually placing each other in check but never taking the other's pieces lest they goad the other player into lunging across the board in order to tear them apart. Clearly, an inability to communicate presents a show-stopping complication to diplomacy.

Why Stop Fighting? Finding the Proper Incentive

If we do find ourselves at war in the future with another spacefaring species, we need to consider what political change would bring the conflict to a close. The distances and time between objectives, battlefields, and conflicts defy modern western strategic military thinking, which is largely predicated on swift strikes, moving faster than the enemy, and in general has produced a human preference for conducting war as bloodless and as fast as possible. While speed will always be important, interstellar distances and the expense of equipping space forces compels us to change our thinking about campaigns.

Once the war is on, finding a reason to stop it may prove more difficult than we thought. Neither the aggressor nor the defender is under any obligation to openly declare their objectives, which makes ending war that much harder. Fighting will be further compounded by mutual hate as the war continues to drag on; for every day that goes by while at war, mutual hate can build up much faster than battles can prove it. We know this will happen because we have already seen it happen here on Earth. John Dower relates how hate and racism can be utilized to energize a war effort in his seminal work *War Without Mercy*.[3] Dower explains as the fighting between the United States and Japan in World War II increased in intensity, and as the stakes of the war rose to include national survival, each nation essentially double-downed on its attempts to paint the other as a subhuman monster which only understood force. Each side's atrocities only fueled these perceptions and made cooperative peace short of capitulation all but impossible.[4]

This situation will certainly persist in space during wars between human forces. However, this condition will get worse if the warring parties are completely different species. Psychologically, mutual hate gives both sides excuses to "desensitize" the other completely, and from a cultural perspective tacitly approves treating the other as a pest or lower life form worthy only of extermination. After all, why bother humanizing a species that isn't even human in the first place? Once a war becomes savage and merciless, it will be only too easy to disregard the adversary's cultural accomplishments and noble virtues, and still easier to forget earlier attempts at peaceful conflict resolution. If the war escalates into a conflict where each side believes their existence as a species is at stake, at that point there is likely no diplomatic force capable of stopping the fighting.

Clearly, this means each side must continue diplomatic efforts and find incentives to cease fighting. To this, we can reliably turn to humanity's conflict resolution record, which despite its spottiness does provide excellent diplomatic examples of peace. In deep space conflict, human wars can expect similar magnanimity; whether or not our understanding and conflict resolution proclivities will extend to wars between humanity and other species remains to be seen.

Treaty Limitations

Ah, the treaty. Terrestrial treaties at once expose both the highest aspirations of humanity's conflict resolution initiatives, and the capriciousness and insincerity of humanity's consistent failure to do as he promises.

Historical treaty adherence rates are, in a word, bad. In earlier times, treaties seldom lasted past the lifetime of the sovereign or government which signed them. In feudal Europe, it was expected that agreements between kings would, unlike their kingdoms, not pass on to their heirs. Governments which possessed ratification processes occasionally disagreed with those in the field who promulgated a treaty, who owing to slow communication and differing passions often abrogated such treaties on the spot, causing mass confusion.

One of the better-known examples of this is the Roman Senate, which occasionally refused to ratify a treaty made by a magistrate or victorious general abroad, thus prolonging a war or conflict. This peculiar form of government often vexed Rome's foes, as was the case in the first treaty attempt during the Second Punic War.[5] Famously, the splendid-isolationist Victorian-era British parliament frequently informed other Great Powers that any treaty promulgated during one government would not last into another.[6]

This track record can create natural cynicism about treaties. In a manuscript discussing the strategic effects and foreign policy implications of atomic weapons for President Eisenhower's review, Frederick Dunn noted treaties of alliance, ostensibly the strongest treaty which exists, have a "decidedly spotty record."[7] Attempting to analyze the significance of America's new superweapon upon foreign policy and national security strategy, Dunn continues by pointing out why treaties usually fail: the subject is too restrictive to one or both parties, forcing them to abandon either their interests or the treaty itself; uncertainties of intention by one or both parties; language difficulties, especially the tendencies for treaties to be ambiguous and noncommittal; and the inability of most treaty drafters to see all possible contingencies and loopholes before concluding the treaty.[8]

Compounding these difficulties is the problem of treaty enforcement across the vast distances of deep space. Here on Earth, adversaries can monitor their opponents' treaty compliance from relatively short distances away, and under the watchful eyes of their agents and third-party friendly nations with whom they share intelligence. Clearly, distances in space will present major enforcement problems to terrestrial states, and these troubles will no doubt be major factors if conflict should break out again.

If we find ourselves in a fight with a non-human civilization, without a third party to sponsor a peace treaty or act as a treaty enforcer, and if distances are so great between our territories as to make treaty enforcement difficult or de facto impossible, then both we and the non-human species are essentially left to our own recognizance whether or not to comply with the treaty. If a peace treaty treads upon one or the other's national interests, the temptation to eventually resume activities contrary to the peace treaty may be too great to resist. Thus, we can expect treaties to present just as much difficulty between human and non-human species as they do only within our own.

Creative treaty enforcement methods could be developed, like stationing distant military units in or near enemy territory or near disputed territory. This of course is expensive and would necessarily be a small contingent who are likely unable to overpower a determined threat to the treaty terms or territory in question, should the adversary break the treaty. Technology will likely offer some solutions, but it is difficult to see just how it could assist the titanic task that is deep space treaty enforcement.

There is yet another question: would a non-human adversary value a treaty in the first place? Clearly, while human nations value treaties and they have served many useful and progressive purposes despite their mottled record, enough of this lackluster history exists to allow a non-human adversary enough political excuses to rationalize their way out of signing a treaty with humanity. They would simply point out that historically humans rarely keep their treaties, cherry pick their choice examples, and claim there would be no reason to make one. A consequence of human historical failures might just be that it comes back to haunt us during diplomacy with a non-human rival. Then again, we should expect a rival to seek any advantage possible during treaty negotiations, which will always include creative—and selective—historical interpretation.

Success or failure of any treaty between humanity and a non-human power would depend largely on the value their culture places on rule of law, truthfulness, and their willingness to do as they say they will do. A legalistic culture would be more willing to approach a treaty with the express purpose of using it as a list to govern behavior between their species and ours, and would probably aim for as lasting a treaty as possible, if for no other reason than to buy time and get rid of us for a while. Still other species may only agree to a simple ceasefire until they can regroup. Above all, the most pressing challenge in future interstellar treaties will be compromising for the lack of trust between two species which have until only recently viewed the other as sub-sentient lower life forms, and who have only just stopped beating each other to death.

The Ease of "Cold War" in Deep Space

Rather than open conflict or pure peace, another likely option exists. Rather than formally commit to a peaceful or warlike relationship, it is possible adversaries who encounter each other in deep space, human or otherwise, will be content to keep a wary but semi-hostile distance from rival forces, and utilize the distance, confusion, and inefficiencies in communication to act bolder then they might if they were nearer to each other. By refusing to declare war or commit to peace, rivals can leverage the relatively slow communication speeds and natural fog of war found in space in order to leave their options open. This could mean engaging in confrontation when they choose, avoidance when they do not, following rival forces at a menacing but safe distance, and engaging in operations up to the threshold of war without actually committing hostile actions. Under the right conditions, rivals are free to pick on enemy or rival vessels or facilities they may find alone or far afield, and to skulk about systems and possessions just out of range of a rival's detection. In general, the distances and communication obstacles present in deep space allow forces which remain aloof and noncommittal the freedom to choose aggression or withdrawal as they see fit, and as long as they are not caught and any action they undertake is below the threshold of war, their behavior could go completely unnoticed or revenged by the victim at a later time. This could be especially true for non-human aggressors, who can always claim miscommunication or misunderstanding as a screen for mischief. As long as a rival's behavior is just shy of the threshold for war, victims may have a difficult time establishing a legal course of action and preferred method of dealing with this strategy.

Obviously, this strategy can work both ways. Any practitioner of Cold War will eventually find themselves subject to the same mischief they give out. Any spacefaring power is just as capable of snooping on a rival as that rival is of snooping on them.

Surprisingly, a Cold War in space is not all bad. Militarily, it could certainly be much worse than icy standoffs in the vastness of space; and while it may feel uncomfortable, Cold War is in fact stable. On the surface there appears to be great waste in maintaining a Cold War standoff with another human political group or non-human spacefaring civilization. By one group refusing to associate with another, each loses the chance for scientific and cultural interchange, technological exchange, and other general activities which please explorers, scientists, and businessmen. Nevertheless, strategists and policymakers should be cautious about upsetting a Cold War situation before noting its benefits and relative stability.

This chapter discussed some diplomatic concerns which come with warfare and thresholds just below war while operating in deep space. While space is unique in its vastness, coldness, and relative mystery, there is little to convince us that the basic nature of war and diplomacy will change as a result of operations which occur in space. This is the same if the adversary is human or non-human; in general, power remains a universal truth which will inform states' and civilizations' decisions and interests, no matter how technologically challenging or how foreboding the distance between these interests happens to be. There is no reason to believe space as a new setting for warfare will appreciably change the political requirement to end wars through diplomacy.

Afterword

> I am of the opinion that no matter how Buck Rogerish things may seem to us now, with the terrific advances made in the art and science of air operations, they should not be overlooked as a possibility for the future.
>
> —General H.H. "Hap" Arnold,
> Commanding General, U.S. Army Air Forces

The United States Air Force has an eye for the future. More than once it has generated sighs and guffaws from sister service officers from across many a meeting table for its seemingly wild technological solutions to problems. Despite this treatment, since its inception the Air Force has provided the answers to complicated technological and security questions demanded by national leaders. When seeking solutions to Soviet nuclear detente, the USAF answered with the Intercontinental Ballistic Missile (ICBM). Faced with the traditional costs and overall dearth of actionable and fresh intelligence during counterinsurgency operations, airmen provided an answer in the form of Remote Piloted Vehicles like the Predator and Global Hawk. Seeking a way around their enemies' air defense systems to collect better intelligence, the USAF in conjunction with the aircraft industry developed the world's fastest reconnaissance aircraft, the SR-71 "Blackbird," capable of outrunning an integrated air defense system's entire compliment of surface-to-air missiles. Once cyber threats became realistic and everyday concerns, it was to the Air Force that national leaders turned to defend the state against cyber threats. Even President Reagan's "Strategic Defense Initiative," while never developed, was to be an Air Force–led initiative, as no other service is capable of the high technology needed for complicated satellite and space operations.

And just as an artilleryman named Giulio Douhet somewhat coherently predicted the onset of future total war waged from the skies, it is not strange when in this text an airman and terrestrial pilot attempts to make somewhat coherent predictions about deep space warfare.

On June 18, 2018, President Donald Trump directed the Department of Defense to develop a sixth branch of the United States Armed Forces: the Space Force. This decision, both hailed and reviled by many different scholars, scientists, military officers, and space enthusiasts, is not as controversial as it first appears to be. Given that space has already been an unofficial battleground for national rivalries here on Earth for many years, establishing a Space Force more resembles the inevitable result of a military operational reality.

Whether or not there should be a completely separate space service, or if space forces should remain subordinated to the USAF, is a very contentious issue. Against the white noise and political maneuvering of this debate, one must take care not to confuse two fundamental issues about space warfare which have yet to be resolved. Essentially, those advocating for an independent Space Force differ in perspective and goals between those advocating for space to be officially considered as a warfighting domain similar to land, sea, and air.

Those that would see space be designated as a separate warfighting domain take an airpower-centric view of space. They see space as simply an extension of the air—a higher high ground. Further, those who are eager to keep spending and political power firmly within the United States Air Force, the current owner of both U.S. military space assets and the air domain, are eager to see space officially designated as a warfighting domain so they may fold the responsibility for war in this domain into the Air Force's portfolio. While this view is understandable, it tends to obscure and discourage future thinking about warfare in space beyond Earth-centric concerns, and also tends to relegate and chain space operations to the air domain—just as land warfare strategists once tried to tie the U.S. Army Air Corps to an exclusive land-support role.

Those wishing for an independent Space Force take a wider and longer-range view of space operations. Generally, those in favor of independence acknowledge that space is a warfighting domain, but not *primarily* a warfighting domain. Rather, independent service supporters favor viewing space akin to the maritime domain here on Earth. Rather than a high ground designed to affect operations on and around Earth only, space is a place to be exploited and controlled, like the sea, rather than owned by any particular hegemon of the moment. These are the idealists and Trekkies; these are the space enthusiasts who think there is more to space than a battleground.

On the surface, it may seem like these two proponents are mutually exclusive. However, synthesis between these two views into a coherent space warfare theory is possible. There is only one seminal event left to catalyze this synthesis: the development of reliable and conventional space-only weaponry designed to function only in space, and for use primarily against space assets. Once these weapons are developed and reliably fielded, a separate space force entity, whether it is a separate service or a corps of the U.S. Air Force, will become inevitable. The responsibility required to safeguard this domain once these weapons proliferate will immediately prove domain specialization is needed, and will likely hit policymakers like a cold bucket of water. The road to an emancipated space force, whatever it looks like, begins there. Until then, the discussion is mostly academic and administrative.

While we do not yet know for sure which form the U.S. approach to space will take, we do know one thing to be true: space combat is coming, as is space mining, exploration, and lunar and extraplanetary settlements. As of this writing, there are only two nations capable of reaching for these goals: The United States, and the People's Republic of China. How these nations currently approach space challenges uniquely reflect their values and the way their governments and societies innovate. China, tempted by the potential for a lunar base and forward spaceport for the potentially lucrative space mining operations to come, has characteristically responded by mobilizing its government through an authoritarian hierarchy to better unify and control the processes necessary to accomplish this goal. The United States, equally aware of space's significance but politically restricted compared to China, has characteristically discouraged barriers to space flight and space entrepreneurship, with an eye toward funding promising private sector innovations and market-based solutions to power the next space revolutions.

Thus, the future of space has two models: government-directed solutions, or market-inspired and government-backed solutions. It is much too early to tell what space will look like fifty years from now, but it is a historical fact that the United States has always responded aggressively and innovatively to security threats in new domains. Its ingenuity is a clear advantage over repressive and authoritarian regimes, especially in the long run. An ambition to treat space as an annex of a state's national programs and interests may seem natural, but in the end will conflict with the U.S. view of space as a solar commons to be used but not dominated. While the United States may start slower out of the blocks, there is no reason to believe they will not rise to the security challenge posed by an overly ambitious Communist police state. Its economic and military power is still immense, despite the musings of many scholars which happily point out its relative decline in power here on Earth. Even though we are seeing the return of a multipolar Great Power world, this fact is largely irrelevant to what the United States can do in space. We are thus left with one conclusion: when it comes to space power, the only thing that can stop the United States is the United States.

Appendix: Useful Formulae

The Drake Equation (number of civilizations)

$$N = R^* f_p n_e f_l f_i f_c L$$

N = the number of civilizations in the Milky Way capable of emitting detectable electromagnetic emissions
R^* = an average of the number of stars born each year (accepted as 1.5)
f_p = the percentage of star systems around which planets form (commonly accepted as 0.9)
n_e = the number of Earthlike planets on average per solar system
f_l = the fraction of these Earthlike planets which actually develop life
f_i = the ratio of planets which birth life and then develop intelligent life
f_c = the percentage possibility of planets with intelligent life developing civilizations
L = likely length of time civilizations will remain detectable before they die out

Force (N)

$$\vec{F} = m\vec{a}$$

m = mass
$\vec{a}$ = acceleration

Universal Gravitation (N)

$$F = G\frac{m_1 m_2}{r^2}, \qquad G = 6.67\times10^{-11}$$

F = Gravitational force
m = mass of two objects (m_1 and m_2 respectively)
G = Universal gravitational constant
r = distances between both objects in question

G-Force / Acceleration (m/s²)

$$g = (v_f - v_i)/(\mathrm{t}_f - \mathrm{t}_i)$$

g = G-Force
v_f = final velocity of an object
v_i = initial velocity of an object
t_f = final time value of a given acceleration
t_i = initial time value of a given acceleration

Glossary

Artificial Intelligence (AI). The ability of machines and technological devices, programs, and creations to sense, process, learn, and adapt to information from their surroundings.

Astrosynchronous. Refers to the relatively stationary and permanent placement of any object in reference to a three-dimensional place in space, rather than to an object.

The Drake Equation. An equation proposed by Cornell astronomer Frank Drake in 1961 what attempts to calculate the number of sentient species which could exist and are potential contacts during the life of our civilization.

Homeworld. The habitable planet where a species first evolves, grows, and develops. To qualify as a homeworld, the species in question must also inhabit the planet or have at one time inhabited the planet. For humanity, this is Earth.

Interplanetary. Any activity or action which takes place between two or more different planets or planetoids *within* the same star system. This term also applies to activity which directly affects two or more planets or planetoids, such as communication or policy.

Interstellar. Any activity or action which takes place *between* two or more different star systems. This term also applies to activity which directly affects two or more star systems, such as communication or policy.

Kardashev Scale. Developed by Soviet astrophysicist Nikolai Kardashev in 1964, the Kardashev Scale classifies species or civilizations into Types 1, 2, or 3 by their ability to harness a portion or all of the energy which reaches their civilization. The scale also addresses the relative technological advancement or knowledge available to their species, with higher types possessing a greater quantity of knowledge and more sophisticated technology.

Sentient. The ability of an organism to perceive and feel things. This definition also implies said organism is capable of rational thought and decision-making.

Xenophilism. An international political philosophy which favors contact and association with foreign entities and organizations.

Xenophobism. An international political philosophy which prefers to avoid contact and association with foreign entities and organizations.

Chapter Notes

Preface

1. John Galsworthy, "Peace of the Air," *The London Times*, 1911, https://www.gutenberg.org/files/57778/57778-h/57778-h.htm.

Introduction

1. Edward S. Miller, *War Plan Orange: The U.S. Strategy to Defeat Japan, 1897–1945* (Annapolis: Naval Institute Press, 1991), 19.
2. Miller, 21.
3. Miller.
4. Miller, 27.
5. Miller, 27.
6. Miller, 27.
7. Miller, 29.
8. Miller, 27.
9. Miller, 4.
10. Miller, 4.
11. Miller, 4.
12. Miller, 4.
13. Miller, 233.
14. Miller, 233.

Chapter 1

1. Travis S. Taylor and Bob Boan, *Alien Invasion: The Ultimate Survival Guide for the Ultimate Attack* (Riverdale, NY: Baen, 2011).
2. For a deeper discussion of the Drake Equation, see Taylor et al. in *Alien Invasion: The Ultimate Survival Guide for the Ultimate Attack*.
3. Cited in Jim Stempel, *The Nature of War: Origins and Evolution of Violent Conflict* (Jefferson, NC: McFarland, 2012), 1.
4. Stempel, 2.
5. Stempel, 7.
6. Joseph Nye and David Welch, *Understanding Global Conflict and Cooperation: An Introduction to Theory and History* (New York: Pearson, 2013), 78.
7. Nye and Welch, 38.
8. "Milestones: 1921–1936—Kellog-Briand Pact," accessed January 19, 2019, https://history.state.gov/milestones/1921-1936/kellogg.
9. "Milestones: 1921–1936—Kellog-Briand Pact."
10. Nye and Welch.
11. "Outer Space Treaty," U.S. Department of State, accessed December 27, 2018, http://www.state.gov/t/isn/5181.htm.
12. Nye and Welch, 4.
13. Nye and Welch, 5.
14. Nye and Welch, 5.
15. Nye and Welch, 3.
16. Nye and Welch, 3.
17. Nye and Welch, 4.
18. Edward C. Mann III, *Thunder and Lightning: Desert Storm and the Airpower Debates* (Maxwell AFB, AL: Air University Press, 1995), 86.
19. Mann, 86.
20. Mann, 86.
21. Mann, 87.
22. Three hundred sextillion is a 3 with 23 zeros.
23. "WMAP Observatory: Mission Overview," accessed October 18, 2018, https://wmap.gsfc.nasa.gov/mission/.
24. Margin of error for this conclusion was 0.4 percent according to NASA's report.
25. Neil deGrasse Tyson, *Astrophysics for People in a Hurry* (New York: W. W. Norton &, 2017).
26. "ESA Gaia Mission: Gaia Creates Richest Star Map of Our Galaxy—and Beyond," accessed September 17, 2018, http://sci.esa.int/gaia/60192-gaia-creates-richest-star-map-of-our-galaxy-and-beyond/.
27. Carl Sagan, "Cosmos: A Personal Voyage" (Cosmos Studios, 2013).
28. Kaiser Wilhelm II, an avid reader of

Mahan and naval enthusiast himself, used to boast that he forced all his naval officers to read Mahan's seminal work *The Influence of Sea Power Upon History*. The Kaiser also made frequent note that he kept a copy of this tome near his bedside.

29. Julian S. Corbett, *Principles of Maritime Strategy*, Dover ed. (Mineola, NY: Dover, 2004).

30. To ease understanding, this is best defined as being an entire generation out-of-date. For example, during operational test of the F-16, it frequently sortied against the F-4 Phantom II, its multi-role predecessor. The F-4, despite having better aircrews and a more-established training and combat legacy, continuously lost to the F-16 in air-to-air engagements due to the latter's vastly better aerial targeting system and maneuverability.

31. Jared Diamond, *Guns, Germs, and Steel: The Fates of Human Societies* (New York: W. W. Norton, 1997), 272.

32. Diamond, 271.

33. Diamond.

34. William Gibson, *Neuromancer* (New York: Ace, 1984).

35. For more information, educator and astrophysicist Neil deGrasse Tyson brilliantly articulates this topic during an interview. See http://bgr.com/2015/12/03/neil-degrasse-tyson-interview-space-exploration/.

Chapter 2

1. Michael Howard, *War in European History*, 1st ed. (New York: Oxford University Press, 2009), 71.

2. Howard, 71.

3. Geoffrey Wawro, *The Franco-Prussian War: The German Conquest of France in 1870–1871* (Cambridge: Cambridge University Press, 2005).

4. Frank J. Allston, *Ready for Sea: The Bicentennial History of the U.S. Navy Supply Corps* (Annapolis: Naval Institute Press, 1995).

5. Allston, 22.

6. Allston, 22.

7. Allston, 129.

8. Allston, 565.

9. Allston, 565.

10. Allston, 23.

11. Allston, 25.

12. Allston, 23.

13. Allston, 151.

14. Allston, 151.

15. Allston, 151.

16. Allston, 151.

17. Tom Wolfe, *The Right Stuff* (New York: Farrar, Straus and Giroux, 1979).

18. Alan Shepard and Deke Slayton, *Moonshot: The Inside Story of America's Race to the Moon* (Atlanta: Turner, 1994).

19. Shepard and Slayton, *Moonshot*.

20. On a more macabre note, it is not immediately clear that planetary conquest would not require complete destruction of the species which originally resided there, if that species was not human. Cohabitation may not be possible for environmental or security reasons, or may be impossible if the war creates a degree of animosity and hatred between our species and our foe which forecloses any possibility of reconciliation.

21. This refers to Air Superiority and Maritime Superiority, military concepts which dictate a combat condition where friendly forces enjoy relative freedom of movement and action in a particular domain. It does not mean total dominance of these domains, which is termed *supremacy* in military jargon.

22. Robert A. Heinlein, *Starship Troopers* (New York: G.P. Putnam's Sons, 1959).

23. Paul K. Davis, *Besieged: 100 Great Sieges from Jericho to Sarajevo* (Oxford: Oxford University Press, 2003).

24. Sun Tzu, *The Art of War*, trans. Samuel B. Griffith (Oxford: Oxford University Press, 1971).

Chapter 3

1. Adrian Goldsworthy, *The Punic Wars* (London: Cassell, 2000).

2. Carl von Clausewitz, *On War*, ed. and trans. Michael Howard and Peter Paret (Princeton: Princeton University Press, 1994), 137.

3. Giulio Douhet, *The Command of The Air*, trans. Dino Ferrari (Washington, D.C.: Air Force History and Museums Program, 1998).

4. Tami Davis Biddle, *Rhetoric and Reality in Air Warfare: The Evolution of British and American Ideas about Strategic Bombing, 1914–1945* (Princeton: Princeton University Press, 2004), 44.

5. Biddle, 45.

6. Clausewitz, 189.

7. Clausewitz, 189.

8. Clausewitz, 189.

9. Clausewitz, 189.

10. Clausewitz, 189.

11. Clausewitz, 189.

12. "Chicxulub Impact Event," accessed Jan-

uary 15, 2019, https://www.lpi.usra.edu/science/kring/Chicxulub/regional-effects/.

13. Incidentally, the chief of the United Nations Office for Outer Space Affairs (UNOOSA) is the official designated ambassador for the United Nations to respond to any contact from extraterrestrial visitors. For more information, see https://www.economist.com/babbage/2010/09/28/the-uns-secretive-alien-ambassador.

14. Thucydides, *The Landmark Thucydides: A Comprehensive Guide to the Peloponnesian War ; with Maps, Annotations, Appendices, and Encyclopedic Index*, ed. Robert B. Strassler, Richard Crawley, and Victor Davis Hanson (New York: Simon & Schuster, 1998).

15. Thucydides.

16. Thucydides.

17. Kenneth B. Pyle, *The Making of Modern Japan*, 2d ed. (Lexington, MA: D.C. Heath and Company, 1996).

18. Pyle, 57.

19. John Findling and Frank Thackeray, eds., *Events That Changed America in the Eighteenth Century* (Westport, CT: Greenwood, 1998), 21.

20. Findling and Thackeray, 22.

21. Other contributing factors to the end of salutary neglect certainly exist, but this reason will suffice for this work. The Seven Years' War is known in the United States as the French and Indian War.

22. Nye and Welch.

23. "The Treaty on European Union and the Treaty on the Functioning of the European Union," Official Journal of the European Union, 2016, https://eur-lex.europa.eu/legal-content/EN/TXT/HTML/?uri=CELEX:C2012/326/01&qid=1401280838965&from=EN.

24. H.G. Wells, "War of the Worlds," 1897.

25. Michael J. Mortlock, *The Egyptian Expeditionary Force in World War I* (Jefferson, NC: McFarland, 2011), 230.

26. Mortlock.

27. E.R. Hooten, *The Greatest Tumult* (London: Brassey, 1991), 14.

28. Harold M. Tanner, *Where Chiang Kai-Shek Lost China: The Liao-Shen Campaign, 1948* (Bloomington: Indiana University Press, 2015), 29.

29. Kenneth Neal Waltz, *Theory of International Politics* (Long Grove, IL: Waveland Press, 2010).

Chapter 4

1. "Space Doctrine," http://www.au.af.mil/au/awc/awcgate/au-18/au180047.htm.

2. This political arrangement was subsequently shattered with President Ronald Reagan's surprise announcement on March 23, 1983, that the United States was planning to develop orbiting satellites equipped with directed energy weapons to shoot down nuclear missiles, ridiculed in the press as "Project Star Wars."

3. Some will no doubt claim the U.S. Air Force's operational tempo following the Cold War, especially from the period of 1993 to present, has prohibited the service from adequately funding space improvement due to other pressing concerns. While this is no doubt in some degree true, space militarization is tightly controlled by national policy vice the U.S. Air Force's own policy, despite its role as the primary space service. It is unlikely the service would have been allowed to innovate and expand space without a renaissance in national space policy, which of this writing has yet to materialize.

4. "Outer Space Treaty."

5. In this chapter, "navy" refers to spacegoing naval forces, and not terrestrial maritime forces.

6. When news reached Rome of Hannibal's action at Cannae, the city was naturally given to riotous panicking and bedlam. Only Rome's sober gravitas—and its prior policy of seeking allies rather than exterminating its neighbors—granted them the political will to continue. Hannibal, finding no local relief for his suffering supply lines, was eventually forced to retreat to Carthaginian territory.

7. "Nelson's Trafalgar Memorandum," accessed December 27, 2018, http://www.bl.uk/learning/timeline/item106127.html.

8. Also known as *wei qi* or *baduk* in Chinese and Korean, respectively.

9. The author's brief experience on a military planning staff seems to confirm this observation.

10. Wawro.

11. Hubertus Strughold, *The Green and Red Planet: A Physiological Study of the Possibility of Life on Mars* (Albuquerque: University of New Mexico Press, 1953).

12. Rick Chen, "Kepler Habitable Zone Planets," Text, NASA, June 16, 2017, http://www.nasa.gov/image-feature/ames/kepler/kepler-habitable-zone-planets.

13. "Open Exoplanet Catalogue," accessed November 10, 2018, http://www.openexoplanetcatalogue.com/.

14. Chen.

15. Mann.

16. Mann, 73.
17. Mann, 73.
18. Mann, 73.
19. Mann, 73.
20. Mann, 73.
21. Mann, 73.
22. Mann, 71.
23. Mann, 74.
24. B.H. Liddell Hart, *Strategy*, 2d rev. ed (New York: Meridian, 1991), 325.
25. Hart, 326.
26. "Air Force Pamphlet 14-210, USAF Intelligence Targeting Guide," 1 February 1998.
27. "Geneva Conventions of 12 August, 1949—UN Documents: Gathering a Body of Global Agreements," accessed January 12, 2019, http://www.un-documents.net/gc.htm.
28. For a deeper look into early twentieth century naval thought, see Robert K. Massie's *Dreadnought: Britain, Germany, and the Coming of the Great War.*
29. John Toland, *The Rising Sun: The Decline and Fall of the Japanese Empire, 1936–1945* (New York: Random House, 1998), 242.
30. The last U.S. Naval frigate was retired in July 2015. The frigate as a class had been systematically ignored by Navy leaders in order to shift resources to other types of vessels, such as guided missile cruisers and AEGIS destroyers. For more information see "End of the 'Ghetto Navy' Is in Sight as Last USN Frigate Cruise Begins," accessed November 11, 2018, https://foxtrotalpha.jalopnik.com/end-of-the-ghetto-navy-is-in-sight-as-last-usn-frigate-1678669074.
31. This is where the term "blackout" and "greyout" originate. As a pilot's brain is deprived of blood and therefore oxygen, first vision degrades, followed by loss of consciousness. The direction of the force is reversed in a high *g* dive, also termed *negative gs*, in which *too much* blood rushes to the brain. This causes the opposite of blackout, "redout," but still inevitably leads to unconsciousness.
32. Actual distance varies completely on the locations of each stellar body in their annual orbits around the sun.
33. Note that "zero gravity" does not actually mean a gravitational value of absolute zero. The various stellar bodies around our carrier, to include the sun, moon, Mars, and others, do in fact exert a gravitational force upon the vessel. For our example, this gravitational force is negligible, but not 0, as this would confound the equations.
34. "Magnetosphere | Science," Solar System Exploration: NASA Science, accessed October 9, 2018, https://solarsystem.nasa.gov/missions/cassini/science/magnetosphere.
35. "Magnetosphere | Science."
36. "Magnetosphere | Science."
37. Clausewitz.

Chapter 5

1. Toland.
2. One striking example is the island of Guam, which has never in its history been successfully defended from invasion. From Spanish colonial powers to the Japanese to the U.S. liberation, Guam's hilly terrain, dense jungles, and relatively few beaches which drive invaders to obvious access points still have not been enough to keep invaders out.
3. Clausewitz.
4. Once mankind reaches the stars, we will be pleasantly surprised at just how much material wealth lies in wait in nearby asteroids and on frozen planetoids silently orbiting our solar system's planets. In the blink of a historical eye, we will probably be just as surprised at how fast our species will come to rely on these extraterrestrial resources.
5. John A. Warden III, "The Enemy as a System," *Airpower Journal* (1995), http://www.au.af.mil/au/afri/aspj/airchronicles/apj/apj95/spr95_files/warden.htm.
6. "History," OPCW, accessed November 11, 2018, https://www.opcw.org/about-us/history.
7. Theodore Gatchel, *At the Water's Edge: Defending Against the Modern Amphibious Assault* (Annapolis: Naval Institute Press, 1996), 123.
8. Gatchel, 122.
9. Gatchel, 123.

Chapter 6

1. Jae-Jung Suh, ed., *Origins of North Korea's Juche* (Lanham, MD: Rowman & Littlefield, 2013), 69.
2. Suh.
3. Suh, 132.
4. Suh, 135.
5. Suh, 139.
6. Suh, 142.
7. "The Kardashev Scale—Type I, II, III, IV & V Civilization," *Futurism*, accessed January 13, 2019, https://futurism.com/the-kardashev-scale-type-i-ii-iii-iv-v-civilization.
8. "The Kardashev Scale—Type I, II, III, IV & V Civilization."

9. "The Kardashev Scale—Type I, II, III, IV & V Civilization."

10. "The Kardashev Scale—Type I, II, III, IV & V Civilization."

11. "The Kardashev Scale—Type I, II, III, IV & V Civilization."

12. Some claim human civilization is much higher than absolute zero. Carl Sagan, for instance, believed humanity had reached 0.7 on the Kardashev scale in 1970, and some believe we are closer to a 0.8 value now.

13. "The Next Really Big Thing: Asteroid Mining Said Worth $700 Quintillion," *World Tribune: Window on the Real World*, accessed October 12, 2018, https://www.worldtribune.com/the-next-really-big-thing-asteroid-mining-said-worth-700-quintillion/.

14. "NEO Basics," accessed February 10, 2019, https://cneos.jpl.nasa.gov/about/neo_groups.html.

15. John S. Lewis, *Asteroid Mining 101: Wealth for the New Space Economy* (Moffett Field, CA: Deep Space Industries, 2015), 101.

16. Matt Williams, "Mars Compared to Earth," *Universe Today* (blog), December 5, 2015, https://www.universetoday.com/22603/mars-compared-to-earth/.

17. Williams.

18. "Asteroids," Solar System Exploration: NASA Science, accessed February 10, 2019, https://solarsystem.nasa.gov/asteroids-comets-and-meteors/asteroids/overview.

19. Lewis, 103.

20. John Wenz, "23 Places We've Found Water in Our Solar System," *Popular Mechanics*, March 16, 2015, https://www.popularmechanics.com/space/a14555/water-worlds-in-our-solar-system/.

21. Lewis.

22. Charles Oman, *Seven Roman Statesmen of the Later Republic: The Gracchi, Sulla, Crassus, Cato, Pompey, Caesar* (HardPress, 2013).

23. Alfred T. Mahan, *The Influence of Sea Power Upon History* (Mineola, NY: Dover, 1987).

24. Mahan.

25. Alfred Mahan, *Mahan on Naval Warfare: Selections from the Writings of Rear Admiral Alfred T. Mahan* (Boston: Little, Brown, 1918).

26. Mahan, *Mahan on Naval Warfare*.

27. Mahan, *The Influence of Sea Power Upon History*.

28. Brent Ziarnick, *Developing National Power in Space* (Jefferson, NC: McFarland, 2015), 21.

29. Ziarnick, 21.

30. Rear Admiral Stephen Bleecker Luce, U.S. Navy, Address Delivered at the U.S. Naval War College 2 June 1903.

Chapter 7

1. John J. Mearsheimer, "Why the Soviets Can't Win Quickly in Europe," *International Security* 7, no. 1, 3–39.

2. Mearsheimer.

3. Michael Shermer, "The Pattern Behind Self-Deception," *TED Talk*, 2010, https://www.ted.com/talks/michael_shermer_the_pattern_behind_self_deception#t-244878.

4. *Star Trek* is a registered trademark of Paramount Pictures registered in the United States. References in this book are used in conjunction with academic fair use legal precedent.

Chapter 8

1. "The Next Really Big Thing," 1.

2. Tyson.

3. "Helix Nebula (NGC 7293): Facts, Location and Images | Constellation Guide," accessed October 12, 2018, http://www.constellation-guide.com/helix-nebula-ngc-7293-caldwell-63-in-aquarius/.

4. Commonly referred to as the tenth commandment, an injunction found in the Hebrew Bible clearly treats wives as though they are property, comparing them to cattle, and inferring they are a resource to be held and protected. This accurately reflects primate behavior regarding sexual rights and territorialism found in other primate species. The exact quote from Exodus 20:17 is: "You shall not covet your neighbor's wife, or his male or female servant, his ox or donkey, or anything that belongs to your neighbor." Similar injunctions appear in Muslim theocratic teachings and Christian cultural traditions which place a wife subservient to her husband.

5. David Biello, "Does Rice Farming Lead to Collectivist Thinking?" *Scientific American*, accessed October 12, 2018, https://www.scientificamerican.com/article/does-rice-farming-lead-to-collectivist-thinking/.

6. Barbara Tuchman, *The Guns of August* (New York: Random House, 1990).

7. Robert K. Massie, *Dreadnought: Britain, Germany, and the Coming of the Great War* (New York: Random House, 1991).

Chapter 9

1. Machiavelli, *The Prince* (Cambridge: Cambridge University Press, 1988).

2. It is unlikely a species can attain the technological sophistication needed to achieve

spaceflight without being able to record information for posterity and to iteratively build upon its scientific discoveries.

3. John Dower, *War Without Mercy* (New York: Random House, 1986).

4. Dower.

5. Goldsworthy.

6. Massie.

7. Bernard Brodie, "The Absolute Weapon: War in the Atomic Age," *The Absolute Weapon: Atomic Power and World Order* (SKIM, 1946), 6.

8. Brodie.

Bibliography

"Air Force Pamphlet 14–210, USAF Intelligence Targeting Guide." 1 February 1998.

Allston, Frank J. *Ready for Sea: The Bicentennial History of the US Navy Supply Corps*. Annapolis: Naval Institute Press, 1995.

"Asteroids." Solar System Exploration: NASA Science. Accessed February 10, 2019. https://solarsystem.nasa.gov/asteroids-comets-and-meteors/asteroids/overview.

Biddle, Tami Davis. *Rhetoric and Reality in Air Warfare: The Evolution of British and American Ideas about Strategic Bombing, 1914–1945*. Princeton: Princeton University Press, 2004.

Biello, David. "Does Rice Farming Lead to Collectivist Thinking?" *Scientific American*. Accessed October 12, 2018. https://www.scientificamerican.com/article/does-rice-farming-lead-to-collectivist-thinking/.

Brodie, Bernard. "The Absolute Weapon: War in the Atomic Age." *The Absolute Weapon: Atomic Power and World Order* (SKIM, 1946).

Chen, Rick. "Kepler Habitable Zone Planets." NASA, June 16, 2017. http://www.nasa.gov/image-feature/ames/kepler/kepler-habitable-zone-planets.

"Chicxulub Impact Event." Accessed January 15, 2019. https://www.lpi.usra.edu/science/kring/Chicxulub/regional-effects/.

Clausewitz, Carl von. *On War*. Edited and translated by Michael Howard and Peter Paret. Princeton: Princeton University Press, 1994.

Corbett, Julian S. *Principles of Maritime Strategy*. Dover ed. Mineola, NY: Dover, 2004.

Davis, Paul K. *Besieged: 100 Great Sieges from Jericho to Sarajevo*. Oxford: Oxford University Press, 2003.

Diamond, Jared. *Guns, Germs, and Steel: The Fates of Human Societies*. New York: W. W. Norton, 1997.

Douhet, Giulio. *The Command of The Air*. Translated by Dino Ferrari. Washington, D.C.: Air Force History and Museums Program, 1998.

Dower, John. *War Without Mercy*. New York: Random House, 1986.

"End of the 'Ghetto Navy' Is in Sight as Last USN Frigate Cruise Begins." Accessed November 11, 2018. https://foxtrotalpha.jalopnik.com/end-of-the-ghetto-navy-is-in-sight-as-last-usn-frigate-1678669074.

"ESA Gaia Mission: Gaia Creates Richest Star Map of Our Galaxy—and Beyond." Accessed September 17, 2018. http://sci.esa.int/gaia/60192-gaia-creates-richest-star-map-of-our-galaxy-and-beyond/.

Findling, John, and Frank Thackeray, eds. *Events That Changed America in the Eighteenth Century*. Westport, CT: Greenwood Press, 1998.

Galsworthy, John. "Peace of the Air." *The London Times*, 1911. https://www.gutenberg.org/files/57778/57778-h/57778-h.htm.

Gatchel, Theodore. *At the Water's Edge—Defending Against the Modern Amphibious Assault*. Annapolis: Naval Institute Press, 1996.

"Geneva Conventions of 12 August, 1949—UN Documents: Gathering a Body of Global Agreements." Accessed January 12, 2019. http://www.un-documents.net/gc.htm.

Gibson, William. *Neuromancer*. New York: Ace, 1984.

Goldsworthy, Adrian. *The Punic Wars*. London: Cassell, 2000.

Hart, B.H. Liddell. *Strategy*. 2d rev. ed. New York: Meridian, 1991.

Heinlein, Robert A. *Starship Troopers*. New York: G.P. Putnam's Sons, 1959.

"Helix Nebula (NGC 7293): Facts, Location and Images | Constellation Guide." Accessed October 12, 2018. http://www.constellation-guide.com/helix-nebula-ngc-7293-caldwell-63-in-aquarius/.

"History." OPCW. Accessed November 11, 2018. https://www.opcw.org/about-us/history.

Hooten, E.R. *The Greatest Tumult*. London: Brassey, 1991.

Howard, Michael. *War in European History*. 1st ed. New York: Oxford University Press, 2009.

"The Kardashev Scale—Type I, II, III, IV & V Civilization." *Futurism*. Accessed January 13, 2019. https://futurism.com/the-kardashev-scale-type-i-ii-iii-iv-v-civilization.

Lewis, John S. *Asteroid Mining 101: Wealth for the New Space Economy*. Moffett Field, CA: Deep Space Industries, 2015.

Machiavelli. *The Prince*. Cambridge: Cambridge University Press, 1988.

"Magnetosphere | Science." Solar System Exploration: NASA Science. Accessed October 9, 2018. https://solarsystem.nasa.gov/missions/cassini/science/magnetosphere.

Mahan, Alfred. *Mahan on Naval Warfare: Selections from the Writings of Rear Admiral Alfred T. Mahan*. Boston: Little, Brown, 1918.

Mahan, Alfred T. *The Influence of Sea Power Upon History*. Mineola, NY: Dover, 1987.

Mann, Edward C., III. *Thunder and Lightning: Desert Storm and the Airpower Debates*. Maxwell AFB, AL: Air University Press, 1995.

Massie, Robert K. *Dreadnought: Britain, Germany, and the Coming of the Great War*. New York: Random House, 1991.

Mearsheimer, John J. "Why the Soviets Can't Win Quickly in Central Europe." *International Security* 7, no. 1 (1982): 3–39. https://doi.org/10.2307/2538686.

"Milestones: 1921–1936—Kellog-Briand Pact." Accessed January 19, 2019. https://history.state.gov/milestones/1921–1936/kellogg.

Miller, Edward S. *War Plan Orange: The US Strategy to Defeat Japan, 1897–1945*. Annapolis: Naval Institute Press, 1991.

Mortlock, Michael J. *The Egyptian Expeditionary Force in World War I*. Jefferson, NC: McFarland, 2011.

"Nelson's Trafalgar Memorandum." Accessed December 27, 2018. http://www.bl.uk/learning/timeline/item106127.html.

"NEO Basics." Accessed February 10, 2019. https://cneos.jpl.nasa.gov/about/neo_groups.html.

"The Next Really Big Thing: Asteroid Mining Said Worth $700 Quintillion." *World Tribune: Window on the Real World*. Accessed October 12, 2018. https://www.worldtribune.com/the-next-really-big-thing-asteroid-mining-said-worth-700-quintillion/.

Nye, Joseph, and David Welch. *Understanding Global Conflict and Cooperation: An Introduction to Theory and History*. New York: Pearson, 2013.

Oman, Charles. *Seven Roman Statesmen of the Later Republic: The Gracchi, Sulla, Crassus, Cato, Pompey, Caesar*. HardPress, 2013.

"Open Exoplanet Catalogue." Accessed November 10, 2018. http://www.openexoplanetcatalogue.com/.

"Outer Space Treaty." U.S. Department of State. Accessed December 27, 2018. http://www.state.gov/t/isn/5181.htm.

Pyle, Kenneth B. *The Making of Modern Japan*. 2d ed. Lexington, MA: D.C. Heath, 1996.

Sagan, Carl. "Cosmos: A Personal Voyage." Cosmos Studios, 2013.

Shepard, Alan, and Deke Slayton. *Moonshot: The Inside Story of America's Race to the Moon*. Atlanta: Turner, 1994.

Stempel, Jim. *The Nature of War: Origins and Evolution of Violent Conflict*. Jefferson, NC: McFarland, 2012.

Strughold, Hubertus. *The Green and Red Planet: A Physiological Study of the Possibility of Life on Mars*. Albuquerque: University of New Mexico Press, 1953.

Suh, Jae-Jung, ed. *Origins of North Korea's Juche*. Lanham, MD: Rowman & Littlefield, 2013.

Sun Tzu. *The Art of War*. Translated by Samuel B. Griffith. Oxford: Oxford University Press, 1971.

Tanner, Harold M. *Where Chiang Kai-Shek Lost China: The Liao-Shen Campaign, 1948*. Bloomington: Indiana University Press, 2015.

Taylor, Travis S., and Bob Boan. *Alien Invasion: The Ultimate Survival Guide for the Ultimate Attack*. Riverdale, NY: Baen, 2011.

Thucydides. *The Landmark Thucydides: A Comprehensive Guide to the Peloponnesian War; with Maps, Annotations, Appendices, and Encyclopedic Index*. Edited by Robert B. Strassler, Richard Crawley, and Victor Davis Hanson. New York: Simon & Schuster, 1998.

Toland, John. *The Rising Sun: The Decline and Fall of the Japanese Empire, 1936–1945*. New York: Random House, 1998.

"The Treaty on European Union and the Treaty on the Functioning of the European Union." *Official Journal of the European Union*, 2016. https://eur-lex.europa.eu/legal-content/EN/TXT/HTML/?uri=CELEX:C2012/326/01&qid=1401280838965&from=EN.

Tuchman, Barbara. *The Guns of August*. New York: Random House, 1990.

Tyson, Neil deGrasse. *Astrophysics for People in a Hurry*. New York: W. W. Norton, 2017.

Waltz, Kenneth Neal. *Theory of International Politics*. Long Grove, IL: Waveland Press, 2010.

Warden, John A., III "The Enemy as a System."

Airpower Journal (1995). http://www.au.af.mil/au/afri/aspj/airchronicles/apj/apj95/spr95_files/warden.htm.

Wawro, Geoffrey. *The Franco-Prussian War: The German Conquest of France in 1870–1871.* Cambridge: Cambridge University Press, 2005.

Wells, H.G. "War of the Worlds," 1897.

Wenz, John. "23 Places We've Found Water in Our Solar System." *Popular Mechanics*, March 16, 2015. https://www.popularmechanics.com/space/a14555/water-worlds-in-our-solar-system/.

Williams, Matt. "Mars Compared to Earth." *Universe Today* (blog), December 5, 2015. https://www.universetoday.com/22603/mars-compared-to-earth/.

"WMAP Observatory: Mission Overview." Accessed October 18, 2018. https://wmap.gsfc.nasa.gov/mission/.

Wolfe, Tom. *The Right Stuff*. New York: Farrar, Straus and Giroux, 1979.

Ziarnick, Brent. *Developing National Power in Space*. Jefferson, NC: McFarland, 2015.

Index